T0366213

Concept Data Analysis

Concept Data Analysis

Theory and Applications

Claudio Carpineto
Fondazione Ugo Bordoni, Italy

Giovanni Romano
Fondazione Ugo Bordoni, Italy

John Wiley & Sons, Ltd

Telephone (+44) 1243 779777

Email (for orders and customer service enquiries): cs-books@wiley.co.uk
Visit our Home Page on www.wileyeurope.com or www.wiley.com

Other Wiley Editorial Offices

John Wiley & Sons Inc., 111 River Street, Hoboken, NJ 07030, USA

Jossey-Bass, 989 Market Street, San Francisco, CA 94103-1741, USA

Wiley-VCH Verlag GmbH, Boschstr. 12, D-69469 Weinheim, Germany

John Wiley & Sons Australia Ltd, 33 Park Road, Milton, Queensland 4064, Australia

John Wiley & Sons (Asia) Pte Ltd, 2 Clementi Loop #02-01, Jin Xing Distripark, Singapore 129809

John Wiley & Sons Canada Ltd, 22 Worcester Road, Etobicoke, Ontario, Canada M9W 1L1

Wiley also publishes its books in a variety of electronic formats. Some content that appears in print may not be available in electronic books.

Library of Congress Cataloging-in-Publication Data

Carpineto, Claudio.
  Concept data analysis : theory and applications / Claudio Carpineto,
Giovanni Romano.
    p. cm.
  Includes bibliographical references and index.
  ISBN 0-470-85055-8 (cloth : alk. paper)
1. Computer science – Mathematics. 2. Machine learning. 3. Information
retrieval. I. Romano, Giovanni. II. Title.
  QA76.9.M35C37 2004
  004′.01′51 – dc22

2004011076

British Library Cataloguing in Publication Data

A catalogue record for this book is available from the British Library

ISBN 0-470-85055-8

Typeset in 11/13pt Palatino by Laserwords Private Limited, Chennai, India.

Contents

Foreword

Concept lattices (*trellis de Galois* in French) is the common name for a specialized form of Hasse diagram that is used in conceptual data processing. Concept lattices are a principled way of representing and visualizing the structure of symbolic data that emerged from Rudolf Wille's efforts to restructure (or modernize) lattice and order theory in the 1980s. Wille's objectives were to recast the presentation of applied lattice and order theory – foundational work in lattice theory from the 1930s by Birkoff but with origins in eighteenth-century mathematics – to better promote communication between lattice theorists and its users. Conceptual data processing (also widely known as 'formal concept analysis') has become a standard technique in data and knowledge processing that has given rise to applications in data visualization, data mining, information retrieval (using ontologies) and knowledge management.

In terms of theory, formal concept analysis has been extended, enabling a powerful general framework for knowledge representation and machine learning called "conceptual knowledge processing". The theoretical developments in the field have proceeded apace, the ideas have an impeccable intellectual pedigree and the scientific literature is vast. *Formal Concept Analysis: Mathematical Foundations* by Bernhard Ganter and Rudolf Wille (Springer-Verlag, 1999), provides a widely read reference to the basic mathematical theory; however, for computer scientists, a broad understanding of the technique and its impact has been slower to emerge. Restructuring the presentation of conceptual data processing relevant to a computer science readership is both sympathetic to Wille's restructuring intentions and is an important achievement of the present volume.

This book covers the basic theoretical and algorithmic foundations of the field and demonstrates their utility for computer scientists interested in information retrieval, data visualization and machine learning. It synthesizes a multitude of sources into a coherent pedagogical presentation that provides an implementation roadmap for practitioners and researchers interested in this important data analysis technique. It can also be used as a textbook for the study of conceptual data processing for undergraduate and postgraduate computer science students. The study of conceptual knowledge processing is a case study in applied computer science. While not alone in this endeavour, I myself have spent the last 7 years of my career studying conceptual data processing. Even so, I gained many surprising and interesting new insights into the field while reading this book.

Modern processors and computer graphics have given increasing scope to the generation and presentation of concept lattices for practical purposes. Since its inception, conceptual data analysis has developed into a research field with an ever increasing number of software tools that facilitate its study. Understanding the basic results of the field through the coherent representation of the basic algorithms (and their complexity) gives scope for the growth and use of the technique and affiliated software. This book condenses 20 years of rapid development in conceptual data processing for applied computer science research into a single source. My belief is that it provides clear evidence of the significance of conceptual data processing to modern applications, and is a model of an emergent interdisciplinary approach to the study of computer science because it synthesizes mathematical, algorithmic and applications ideas. This is a book that will contribute to the study of the conceptual data processing in computer and information science departments throughout the world and is highly recommended!

Peter Eklund

Preface

Motivation and contents

The recent advances in the acquisition, storage and transmission of data in digital format, as witnessed by the advent of the World Wide Web, have dramatically increased the need for tools that effectively support users in retrieving, understanding and mining the information and knowledge contained in such data. Concept data analysis may play an important role and help fill this gap thanks to its simplicity, elegance and versatility.

Concept data analysis differs from statistical data analysis in that the emphasis is on recognizing and generalizing structural similarities, such as set inclusion relation from the data description, and not on mathematical manipulations of probability distributions. In essence, by using concept data analysis one can turn any collection of objects described by a set of properties into a lattice of concepts, where each concept covers a subset of the objects contained in the given collection and is described by the properties shared by the objects pertaining to that concept. Such a conceptual representation, termed a 'concept lattice', can be extracted from different or heterogeneous types of data (e.g., text, semistructured or structured data) and can then be used to support various kinds of content management tasks.

Since Rudolf Wille's seminal paper in 1982, concept data analysis has attracted a growing number of researchers and practitioners. The field has evolved into a well-recognized research community, and it is now attempting to gain more widespread acceptance by broadening its application base. This effort has been favoured by the new Web programming paradigm, which has greatly facilitated the construction and the utilization of the graphical interfaces that are needed for visualizing and manipulating concept lattices, ensuring simplicity, efficiency, portability and adaptability.

The general impression is that concept data analysis represents a powerful technology for content processing, the potentials of which have not been fully exploited to date. Symmetrically, there seem to be plenty of interesting applications that could benefit from the kind of content processing performed by concept data analysis. One of the primary goals of this book is to make this link more visible and concrete.

The theoretical foundations of concept data analysis are now well understood. The core theory has remained substantially stable and is presented in a number of papers. Furthermore, a book has recently been published that provides a thorough treatment of the mathematical features of concept lattices. As the theoretical side is well covered in the literature, this book provides an introduction to the issue, focusing on the main ideas which are necessary for developing algorithms and building applications.

Besides theory, a number of advances have also been reported on the methodological side, especially over the last few years. There are excellent papers on specific algorithms that cover some aspects of concept data analysis, such as the construction and visualization of concept lattices. But, in general, there is no single collection of papers, let alone a book, that provides a synthetic, comprehensive treatment of the full range of algorithms available for concept data analysis, spanning the creation, maintenance, display and manipulation of concept lattices. This book is a first attempt to fulfil this need.

Equipped with sound theoretical machinery and with a toolbox for implementing it, the next step is to look for problems that can be solved by the new technology. This aspect has been somewhat overlooked to date, which may have hindered the diffusion and popularity of the overall methodology. The application side is the main thrust of this book, with several innovative and fully worked proposals. The title of the book – *Concept Data Analysis* – differs from the heading under which most recent theoretical work in the field has been done – *Formal Concept Analysis* – because it intends to highlight the shift from a mathematics to a computer science perspective.

Previous research has mostly focused on a relatively straightforward application of the basic techniques of concept data analysis in different domains (e.g., medicine, psychology, software). This book takes a somewhat broader perspective by focusing on application areas which cut across multiple domains.

Three main areas are explored: information retrieval from textual data, interactive mining of documents or collections of documents (including Web documents), and rule mining from structured data. It is shown that by using the contextual information made available by concept data analysis techniques – possibly combined with other techniques for information processing and management such as those developed in information retrieval, data mining, human–computer interaction, and Web-based information systems – one can build more flexible, precise or efficient systems.

The potential of concept data analysis for text mining is especially noteworthy, as illustrated by two detailed case studies, which are briefly explained below.

The ACM Digital Library, the subject of the first case study, illustrates a two-step methodology for mining collections of documents. In a first step, conceptual data analysis is used to cluster ACM's publications based on their indexes and on the ACM classification system; subsequently, a user interface visualizes the resulting clustering structure and lets users query, browse, or reorganize it based on their personal needs. The features of the system are well beyond the mining capabilities of ACM Digital Library's front end, as well as of those of most on-line digital libraries.

The second case study is concerned with a similar subject – interactive mining of documents – in a different domain, namely Web-based retrieval. The lack of effectiveness of the interfaces of current search engines is addressed by using conceptual data analysis to help users interpret Web retrieval results and select those of interest. The case study focuses on the construction of a Web-based client-server system, called CREDO, that allows users to input a query to available search engines and navigate the concept lattice of the results returned by the engines. Mining Internet results using concept lattices presents additional difficulties over mining intranet collections because of computational constraints. The main steps that are necessary to build the system are illustrated in detail, and its potential is further demonstrated through an implementation of an on-line version available at:

http://credo.fub.it/

CREDO may be of interest to anybody who needs to search the Web. It is especially useful for grouping the results of an ambiguous query according to the meanings of the query, or for getting a quick idea about the contents of the documents that reference a given entity.

Utility of the book

This is the first synthetic and comprehensive view of the methods developed for concept data analysis. This is also the first time that concept data analysis techniques have been seen as a component for performing more complex performance tasks which require adaptation to and/or integration with different data representations and information processing techniques. Thus, the book allows readers to solve problems in practice using concept data analysis.

Using the information in this book, readers will be able to build a conceptual representation of any database of interest (e.g., specialized digital libraries, Web documents returned by a search engine, scientific or social records, annotated multimedia archives, e-mail messages) and to use the representation found for performing a number of tasks involving explicit management of the original database content (e.g., document mining, contextual ranking, rule discovery).

The accompanying system, CREDO, clearly demonstrates the usefulness of this approach for the very common task of searching the Web. It will automatically cluster the retrieved documents into a lattice of meaningful concepts and allow the user to browse through such concepts to find the documents of interest quickly or discover unknown information.

Prerequisites

The material presented in the book is accessible to all readers with a basic knowledge of data structures, algorithms and programming. The introduction to the theoretical foundations of concept data analysis is self-explanatory. The mathematical notation is simple, symbols are kept to a minimum, and no prior knowledge about order theory is assumed on the part of the reader. Similarly, no background knowledge of the application areas being investigated is required by the reader.

Readership

The book is primarily intended for researchers in information processing and management – including various fields such as data analysis, information systems, information retrieval, digital libraries, and data mining – as well as for industry practitioners who are interested in creating a commercial product for concept data analysis or developing content management applications in any of the areas analysed in the book.

The book is also of value to senior undergraduate or postgraduate students, as a reference book or secondary text for courses on information science, information management, information retrieval, artificial intelligence, and data mining. It can be used to complement more conventional subjects or for developing multidisciplinary projects involving databases, text processing, and Web information systems.

Overall, we hope that the book will contribute to disseminating the study of concept data analysis as well as to increasing the practice of this discipline in computer science departments and related industries.

Organization of the book

The book consists of two parts: one on theory and algorithms followed by one on applications.

Part I contains two chapters. Chapter 1 covers basic elements of order theory, then introduces the notions of formal concept and concept lattice for single-valued contexts, and finally discusses how to deal with many-valued contexts.

Chapter 2 describes the techniques available for constructing, updating and visualizing concept lattices, as well as for enhancing them with knowledge not encoded in the original context. The algorithms for lattice construction are distinguished according to the output produced, i.e., the set of concepts without the edges, the concept lattice, or a partial concept lattice. The algorithms for updating lattices are broken down into two classes, those for incremental construction of concept lattice as new objects are added (which can also be used on a static set of objects), and those for updating lattices in response to general variations in the context table. The techniques described for visualizing concept lattices include hierarchical folders, nested line diagrams, and focus+context views. Finally, two types of external knowledge are considered for inclusion in a concept lattice, i.e., subsumption hierarchies (whether structuring single many-valued attributes or all one-valued attributes) and user constraints.

Part II shows how all of the preceding principles and algorithms fit together when applied to a range of text and data analysis problems. It is divided into three chapters. Chapter 3 covers interactive query modification and automatic text ranking. Interactive query modification can be performed either by mapping the user query onto the document lattice and then navigating around such a query concept, or by incorporating a thesaurus into the document lattice and then using the thesaurus hierarchies to query the enhanced lattice. Automatic text ranking exploits the interdocument similarities detected by concept data analysis for alleviating the well-known vocabulary problem encountered by conventional information retrieval systems, i.e., the difficulty of dealing with word mismatch between query and documents. The issue of automatically generating a set of index terms for constructing the concept lattice of a collection of unindexed documents is also dealt with in Chapter 3.

Chapter 4 covers the two case studies introduced above. The ACM Digital Library case study is discussed first. Following an illustration of the current retrieval interface to the ACM Digital Library, we discuss its lack of provision for mining the content of the available documents, and then go on to describe a concept lattice-based framework which allows hybrid *retrieval and mining* strategies. The second part of the chapter deals with mining Web retrieval results. After motivating the use of concept lattices for solving this task, we describe the design and implementation of CREDO and present example sessions of the interaction between the user and the system.

Chapter 5 focuses on the discovery from structured data of multiple types of data dependencies that are logically related such as implications, functional dependencies, association rules and classification rules. Using the concept lattice built from the input database as an intermediate representation results in methods that are more efficient or produce more accurate or compact results than most conventional rule miners.

At the end of each chapter, references to earlier works on which some of the material of the book is based are provided. A discussion of research issues and trends, and a selected bibliography are also included.

Omissions

A single book can hardly treat all the dimensions of the background and methodology of concept data analysis and also embrace all practical developments in the area. We have chosen to limit the theoretical foundations to basic concepts, as there are specialized texts for more advanced subjects. For the algorithms, all the main tasks are discussed and the most significant approaches are represented, although there may well be some variants or recent improvements that are not covered by this book. The application side is more problematic, partly because the spectrum of data analysis domains where concept data analysis has proven its value has been constantly expanding, and partly because there are some areas in which we are not expert. Among the omitted applications, it is worth mentioning that substantial research and applications have been developed and reported for software engineering, object-oriented languages, and natural language processing. Some of these approaches will be indicated as relevant throughout the book.

Web page

Information relevant to this book will be maintained at:

http://search.fub.it/cda

The Web page contains the Table of Contents, Foreword, Preface, and Acknowledgements, and the e-mail address of the contact authors. Any errors discovered in this book will be posted on this page. The Web page also contains a pointer to the CREDO system.

Acknowledgements

We are especially indebted to Peter Eklund, who encouraged us from the very beginning of this project and has provided assistance and invaluable feedback at various stages. We thank Sergei Kuznetsov and Gerd Stumme for their thorough reviews of Part I and Chapter 5, respectively, and Catherine Roberts for helping with the English. We would also like to thank three anonymous reviewers who read an early proposal for this book and provided us with good feedback and insight. We express our gratitude to Laura Kempster and the other members of the John Wiley & Sons, Ltd. team for their interest and patience with many delays in the schedule and for their production work. Last but not least, we thank the support of our institution, Fondazione Ugo Bordoni, without which this book would not have been written. The first author would finally like to thank Cristina and Livia for their loving support and endless patience.

Part I

Theory and Algorithms

1

Theoretical Foundations

In this chapter we offer a condensed treatment of the theoretical foundations of concept data analysis. We cover only the basic notions necessary from an application point of view; more advanced theoretical features can be found in the publications referenced at the end of the chapter.

We start by presenting the rudiments of ordered sets and intersection structures, providing a foundation on which to build the two subsequent sections.

The rest of the chapter serves as an introduction to the theory of concept lattices for computer scientists. We formalize the notions of concept, context, and concept lattice, and then discuss how to extend the basic theory to deal with more complex data types.

1.1 Basic notions of orders and lattices

A binary relation $\leq$ on a set P is called an *order relation* (or *partial order relation*) on P if, for all $x, y, z \in P$:

1. $x \leq x$,
2. $x \leq y$ and $y \leq x$ imply $x = y$,
3. $x \leq y$ and $y \leq z$ imply $x \leq z$.

These conditions are referred to, respectively, as reflexivity, antisymmetry, and transitivity. A set P equipped with an order relation $\leq$, denoted by $(P, \leq)$ or just by P, is called an *ordered set* (or *partially ordered set* or simply *order*).

Examples of ordered sets are the power-set $\wp(X)$ of all subsets of any set X with set inclusion and the real numbers $\mathcal{R}$ with the usual $\leq$ relation. The latter ordered set is a *chain*, because for all $x, y \in \mathcal{R}$, either $x \leq y$ or $y \leq x$ (i.e., any two elements of $\mathcal{R}$ are comparable).

Concept Data Analysis: Theory and Applications. Claudio Carpineto and Giovanni Romano
 ISBN: 0-470-85055-8

Let P be an ordered set and let $x, y, \in P$. Element x is *covered* by y (or y *covers* x) if

$x < y$ and there is no $z \in P$ such that $x < z < y$.

We write $x \prec y$. We can also say that x is a *lower neighbour* of y, or that y is an *upper neighbour* of x. If P is finite, $x < y$ if and only if there exists a finite sequence of covering relations between x and y.

Every finite ordered set $(P, \leq)$ can be drawn. We use small circles to represent the elements of P and interconnecting lines to indicate the covering relation. The drawing is such that element x is placed below element y if $x \prec y$. This representation is called a *line diagram* or *Hasse diagram*.

Clearly, the same ordered set may have many different diagrams, and the notion of a good line diagram is difficult to formalize. In Figure 1.1 we show several possible ways to draw the same ordered set.

From a line diagram we can easily tell whether one element is less than another: $x < y$ if and only if there is an ascending path from x to y. Figure 1.2 presents line diagrams for all ordered sets with three elements.

Let P and Q be two ordered sets. A map $\varphi: P \to Q$ is called *order-preserving* if,

for all $x, y \in P, x \leq y$ in P implies $\varphi(x) \leq \varphi(y)$ in Q.

If φ is an order-preserving map such that

$\varphi(x) \leq \varphi(y)$ implies $x \leq y$,

then φ is called an *order-embedding* map. A bijective order-embedding is called an *(order-)isomorphism*.

Figure 1.3 shows two maps between ordered sets. The map φ_1 is order-preserving, but not order-embedding, because $\varphi_1(b) \leq \varphi_1(c)$ in Q whereas b

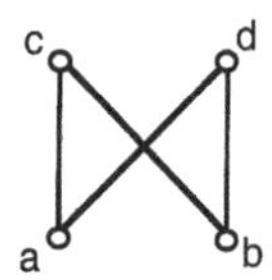

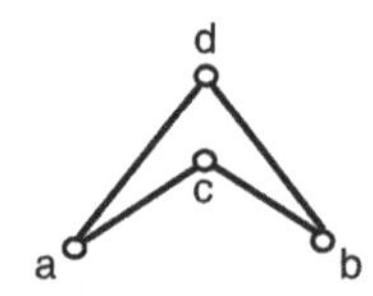

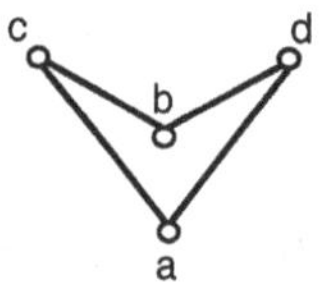

Figure 1.1 Alternative line diagrams for the ordered set $P = \{a, b, c, d\}$, with $a < c$, $a < d$, $b < c$, and $b < d$.

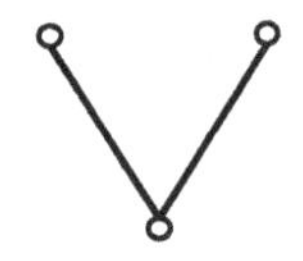
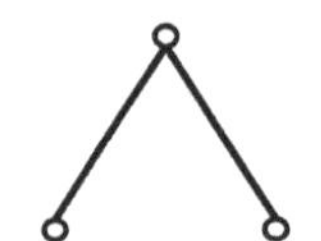

Figure 1.2 Line diagrams of all possible ordered sets with three elements.

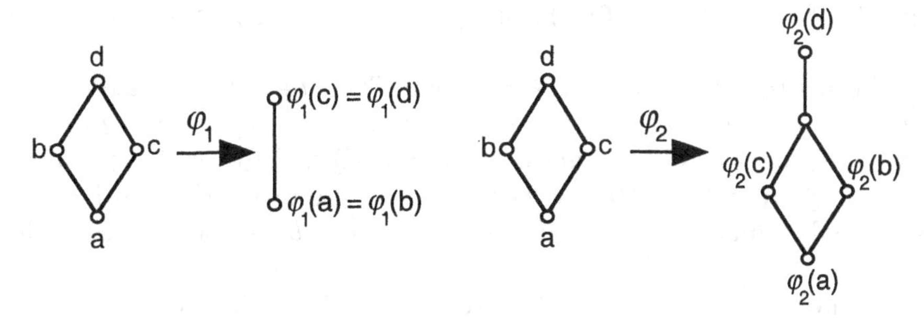

Figure 1.3 Examples of maps between ordered sets.

and c are incomparable in P. The map φ_2 is an order-embedding, but not an order-isomorphism, because a new element is introduced between $\varphi_2(d)$ and $\varphi_2(c)$ (or $\varphi_2(b)$).

We now introduce the duality principle. Given any ordered set $(P; \leq)$, the *dual* of $(P; \leq)$ is defined as $(P; \geq)$, where the order relation $\geq$ is the inverse of $\leq$ (i.e., $y \geq x$ holds if and only if $x \leq y$). A line diagram for $(P; \geq)$ can be obtained simply by turning a diagram for $(P; \leq)$ upside down.

The dual statement of an order-theoretic statement can be derived by replacing the symbol $\leq$ by $\geq$. A statement holds for an ordered set if and only if the dual statement holds for the dual set. This is called the *duality principle.*

For instance, given the statement 'there are two elements greater than both $\varphi(b)$ and $\varphi(c)$' holding in the far right-hand diagram in Figure 1.3, the dual statement 'there are two elements smaller than both $\varphi(b)$ and $\varphi(c)$' holds in the dual of the same diagram. The duality principle can often be used to simplify theorem proofs.

Let P be an ordered set and $Q \subseteq P$. Q is a *down-set* or *order ideal* if $x \in Q$ and $y \leq x$ always imply $y \in Q$. Dually, Q is an *up-set* (or *order filter*) if $x \in Q$ and $x \leq y$ always imply $y \in Q$. For any subset Q of P, we define

$\downarrow Q = \{y \in P \mid (\exists x \in Q)\ y \leq x\}$, read 'down Q', and

$\uparrow Q = \{y \in P \mid (\exists x \in Q)\ x \leq y\}$, read 'up Q'.

Similar definitions hold for any $x \in P$:

$\downarrow x = \{y \in P \mid y \leq x\}$ and

$\uparrow x = \{y \in P \mid x \leq y\}$.

The set $\downarrow Q$ is the smallest down-set containing Q and Q is a down-set if and only if $Q = \downarrow Q$; dual statements hold for $\uparrow Q$. Consider the ordered set in Figure 1.1. The set $\{a, b, c\}$ is a down-set, but not an up-set (because $\uparrow a = \{c, d\}$). The set $\{a, c, d\}$ is an up-set, but not a down-set. The set of all down-sets of P, denoted by $\mathcal{O}(P)$, is itself an ordered set, under the inclusion relation.

Let P be an ordered set and Q a subset of P. Then $a \in Q$ is a *maximal* element of Q if $a \leq x \in Q$ implies $a = x$. If $x \leq a$ for every $x \in Q$, a is the *greatest* (or *maximum*) element of Q. A *minimal* element of Q and the *least* (or *minimum*) element of Q are defined dually. For example, the ordered set in Figure 1.1 has maximal elements c, d and minimal elements a, b, but no greatest or least element.

The greatest element of P, if it exists, is called the *top* element of P ($\top$); the least element of P, if it exists, is called the *bottom* element of P ($\perp$). For instance, d and a are, respectively, the top and the bottom elements of the left ordered set in Figure 1.3.

New ordered sets can be built from existing ones in several ways. Let $(P_1; \leq)$ and $(P_2; \leq)$ be disjoint ordered sets. The *cardinal sum* (or *disjoint union*) of $(P_1; \leq)$ and $(P_2; \leq)$ is the ordered set $(P_1 \cup P_2; \leq)$, the order relation being defined as follows: $x \leq y$ if and only if either $x, y \in P_1$ and $x \leq y$ in P_1 or $x, y \in P_2$ and $x \leq y$ in P_2. A line diagram for the cardinal sum can be obtained by placing side by side diagrams for the summands.

The *linear sum* of $(P_1; \leq)$ and $(P_2; \leq)$ is defined as the cardinal sum except that $x \leq y$ holds also if $x \in P_1, y \in P_2$. A diagram for the linear sum is formed by placing a diagram for P_2 directly above a diagram for P_1 and then adding a line segment from each maximal element of P_1 to each minimal element of P_2.

The (*direct*) *product* of two (disjoint) ordered sets $(P_1; \leq)$ and $(P_2; \leq)$ is the ordered set $(P_1 \times P_2; \leq)$, where $(P_1 \times P_2)$ is the Cartesian product of P_1 and P_2 and the order relation on the product is such that

$$(x_1, x_2) \leq (y_1, y_2) \longleftrightarrow x_1 \leq y_1 \text{ and } x_2 \leq y_2.$$

An example of a product of two ordered sets is shown in Figure 1.4. The definition of the product, as well as of the sums introduced earlier, can be extended to any number of ordered sets. These composition operators can be

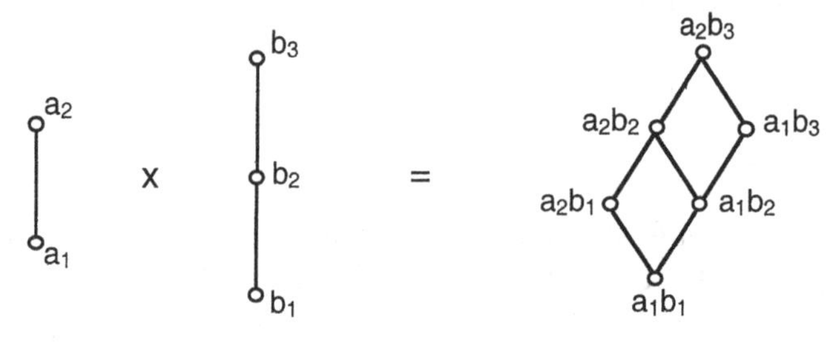

Figure 1.4 Product of two ordered sets.

useful not only to join two or more ordered sets, but also to analyse a given ordered set as if it consisted of simpler parts.

Now we introduce the notions of a lattice and a complete lattice, which are two important classes of ordered sets characterized by the existence of certain upper bounds or lower bounds of their elements.

Let $(P; \leq)$ be an ordered set and S a subset of P. An element $x \in P$ is called an *upper bound* of S if $s \leq x$ for all $s \in S$. A *lower bound* is defined dually. The set of all upper bounds of S is denoted by S^u and the set of all lower bounds by S^l; S^u is always an up-set and S^l a down-set.

If there is a least element S^u, it is called the *least upper bound* or *supremum* of S, denoted by sup S. Dually, a greatest element in S^l, if it exists, is called the *greatest lower bound* or *infimum* of S, denoted by inf S.

Supremum and infimum are frequently also called *join* and *meet*. We write $x \vee y$ ('x join y') instead of sup $\{x, y\}$ and $x \wedge y$ ('x meet y') instead of inf $\{x, y\}$. We also write $\bigvee S$ (the 'join of S') and $\bigwedge S$ (the 'meet of S') instead of sup S and inf S.

In an ordered set P, $x \vee y$ may fail to exist either because x and y have no upper bound or because they have no least upper bound. In Figure 1.1, $c \vee d$ does not exist because there is no common upper bound, while $a \vee b$ does not exist as the set of upper bounds ($\{c, d\}$) does not have a least element.

An ordered set $(P; \leq)$ is called a *lattice* if for any pair of elements x and y in P the supremum $x \vee y$ and the infimum $x \wedge y$ always exist. An ordered set $(P; \leq)$ is called a *complete lattice* if the supremum $\bigvee S$ and the infimum $\bigwedge S$ exist for any subset S of P. Every complete lattice has a top element, also called the *unit* element of the lattice, and a bottom element, called the *zero* element.

A complete lattice can be characterized in other ways. Two useful propositions are that every non-empty finite lattice is complete and that a non-empty ordered set in which the infimum exists for every subset is a complete lattice.

For example, the ordered set in Figure 1.1 is not a lattice. The set $\Re$ is a lattice under its usual order ($x \vee y = \max\{x, y\}$ and $x \wedge y = \min\{x, y\}$), but it is not complete because it lacks a greatest and a least element. Each of the ordered sets in Figure 1.4 is a complete lattice. The ordered power-set $(\wp(X); \subseteq)$ is a complete lattice in which the lattice operations join and meet are set union and intersection. The unit element is X and the zero element is $\varnothing$.

A complete lattice L is called *distributive* if the following distributive laws hold for all $x, y, z \in L$:

$$x \wedge (y \vee z) = (x \wedge y) \vee (x \wedge z)$$

$$x \vee (y \wedge z) = (x \vee y) \wedge (x \vee z)$$

An ordered set $(P; \leq)$ is called a *join semilattice* (*meet semilattice*) if only the supremum (infimum) exists for any subset S of P.

A non-empty subset N of a lattice L is called a *sublattice* of L if

$$a, b \in M \Rightarrow a \vee b \in M \text{ and } a \wedge b \in M.$$

For example, the subset $\{a_1b_1, a_1b_2, a_2b_2, a_2b_3\}$ is a sublattice of the product lattice in Figure 1.4, while the subset $\{a_1b_1, a_1b_2, a_2b_1, a_2b_3\}$ is not a sublattice of the same lattice.

It would be useful to have a subset of distinguished elements of L by which to represent all the elements of L. We will now address this aspect. Let L be a lattice. An element $x \in L$ is *join-irreducible* (or $\bigvee$*-irreducible*) if

$$a < x \text{ and } b < x \text{ imply } a \vee b < x \text{ for all } a, b \in L,$$

that is, if it cannot be represented as the supremum of strictly smaller elements. A *meet-irreducible* (or $\bigwedge$*-irreducible*) element is defined dually. We denote the set of join-irreducible elements of L by $\mathcal{J}(L)$ and the set of meet-irreducible elements of L by $\mathcal{M}(L)$.

The upper neighbours of the zero element (called *atoms*, if they exist) are always $\bigvee$-irreducible, while the lower neighbours of the unit element (called *coatoms*) are always $\bigwedge$-irreducible.

In a chain, every non-zero element is join-irreducible (thus, if L is an n-element chain, $\mathcal{J}(L)$ is an $(n-1)$-element chain). If L is the product of L_1 and L_2, the element (x_1, x_2) is not join-irreducible unless either x_1 or x_2 is zero. For the product in Figure 1.4, for example, the only two join-irreducible elements are (a_2, b_1) and (a_1, b_2).

In a finite lattice, an element is join-irreducible if and only if it has exactly one lower cover (or lower neighbour). Dually, an element is meet-irreducible if and only if it has exactly one upper cover (or upper neighbour). This property makes it easy to find $\mathcal{J}(L)$ or $\mathcal{M}(L)$ from a diagram of L.

A set $P \subseteq L$ is called *join-* or *supremum-dense* in L if for every $a \in L$ there exists $A \subseteq P$ such that $a = \bigvee A$; dually, it is *meet-* or *infimum-dense* if for every $a \in L$ there exists $A \subseteq P$ such that $a = \bigwedge A$.

Let L be a finite lattice. Every supremum-dense subset in L contains all join-irreducible elements and every infimum-dense subset contains all meet-irreducible elements. Conversely, the set $\mathcal{J}(L)$ is supremum-dense and $\mathcal{M}(L)$ is infimum-dense. Thus, each element of L is the join of join-irreducible and the meet of meet irreducible elements. In particular,

$$a = \bigvee \{x \in \mathcal{J}(L) \mid x \leq a\} \text{ for all } a \in L, \text{ and}$$

$$a = \bigwedge \{x \in \mathcal{M}(L) \mid a \leq x\} \text{ for all } a \in L.$$

We now introduce the notions of closure system and closure operator. Let H be a set and let $\mathcal{K}$ be a set of subsets of H. $\mathcal{K}$ is a *closure system* on H if

$\bigcap_{i\in I} A_i \in \mathcal{K}$ for every non-empty subset $A_{i\in I} \subseteq \mathcal{K}$ and

$\mathcal{K} \in H$.

In other words, $\mathcal{K}$ must be closed under intersection and topped. The power-set is clearly a closure system. Other examples of a closure system are the subspaces of a vector space, the subgroups of a group, and the equivalence relations on a set.

If $\mathcal{K}$ is a closure system, then $(\mathcal{K}; \subseteq)$ is a complete lattice in which

$$\bigwedge \{A_i \mid i \in I\} = \bigcap_{i\in I} A_i,$$

$$\bigvee \{A_i \mid i \in I\} = \bigcap \{B \in \mathcal{K} \mid \bigcup_{i\in I} A_i \in B\}.$$

A *closure operator* on H is a map $\varphi : \wp(H) \rightarrow \wp(H)$ such that, for all $A, B \subseteq H$:

1. $A \subseteq \varphi(A)$,
2. if $A \subseteq B$, then $\varphi(A) \subseteq \varphi(B)$,
3. $\varphi(\varphi(A)) = \varphi(A)$.

An example of closure operator is the map $\downarrow P$ introduced earlier. The corresponding closure system is the set $\mathcal{O}(P)$ of all down-sets of P.

Closure systems and closure operators are closely related. The set of all closures of a closure operator is a closure system and it forms a complete lattice, when ordered by inclusion, in which

$$\bigwedge \{A_i \mid i \in I\} = \bigcap_{i\in I} A_i,$$

$$\bigvee \{A_i \mid i \in I\} = \varphi\big(\bigcup_{i\in I} A_i\big).$$

Conversely, given a closure system $\mathcal{K}$ on H, a closure operator $\varphi_{\mathcal{K}}$ on H is defined by

$$\varphi_{\mathcal{K}}(A) := \bigcap \{B \in \mathcal{K} \mid A \subseteq B\}.$$

As every closure system $\mathcal{K}$ can be seen as the set of all closures of a closure operator, the elements of $\mathcal{K}$ are called closures too. Every complete lattice is isomorphic to the lattice of all closures of a closure system.

Let P and Q be ordered sets. A pair of maps

$$\varphi : P \longrightarrow Q \text{ and } \psi : Q \longrightarrow P$$

is called a *Galois connection* between P and Q if:

1. $p_1 \leq p_2 \Rightarrow \varphi p_1 \geq \varphi p_2$,
2. $q_1 \leq q_2 \Rightarrow \psi q_1 \geq \psi q_2$,
3. $p \leq \psi\varphi p$ and $q \leq \varphi\psi q$

or, equivalently, if

$$p \leq \psi q \Leftrightarrow q \leq \varphi p.$$

In the case of a Galois connection between $\wp(P)$ and $\wp(Q)$, the map $\psi\varphi$ is a closure operator on P and the map $\varphi\psi$ is a closure operator on Q.

The last issue of this brief introduction to orders and lattices concerns the embedding of an ordered set into a complete lattice. Let P be an ordered set. If $\varphi : P \rightarrow L$ and L is a complete lattice, then L is called a *completion* of P, via the order-embedding φ. In general, there are several order-embeddings of an ordered set into a complete lattice. An example is the set $\mathcal{O}(P)$ of all down-sets of P, although this is not a very compact completion.

The best known embedding is the *Dedekind–MacNeille completion*, defined as follows:

$$DM(P) := \{A \subseteq P | A^{ul} = A\}.$$

This is the smallest complete lattice in which P can be embedded.

1.2 Context, concept, and concept lattice

We start by formalizing the notion of context.

A *context* is a triple (G, M, I) consisting of two sets G and M and a relation I between G and M. The elements of G are called the *objects* and the elements of M are called the *attributes* (the letters G and M come from the German *Gegenstände* and *MerKmale*). The relation I is also called the *incidence relation* of the context. We write gIm or $(g, m) \in I$ to mean that the object g has the attribute m. We may think of the set of attributes associated with an object as a bit vector; each bit corresponds to a possible attribute and is on or off depending upon whether an object has that attribute. Throughout this book we will usually assume that G and M are finite.

In Table 1.1 we show a (very limited) context for vertebrate animals, where the objects are animals and the attributes are properties describing the animals. A cross in the ij position of the table indicates that object i is described by attribute j, or, equivalently, that attribute j describes object i. For instance, a bat can fly, has a skeleton, has wings, and is viviparous, while having wings is shared by bats, eagles and penguins.

Table 1.1 A context for vertebrate animals

		breathes in water (a)	can fly (b)	has beak (c)	has hands (d)	has skeleton (e)	has wings (f)	lives in water (g)	is viviparous (h)	produces light (i)
1	Bat		×			×	×		×	
2	Eagle		×	×		×	×			
3	Monkey				×	×			×	
4	Parrot fish	×		×		×		×		
5	Penguin			×		×	×	×		
6	Shark	×				×		×		
7	Lantern fish	×				×		×		×

For a set $A \subseteq G$ of objects define

$$A' = \{m \in M \mid gIm \text{ for all } g \in A\}.$$

Correspondingly, for a set $B \subseteq M$ of attributes define

$$B' = \{g \in G \mid gIm \text{ for all } m \in B\}.$$

In other words, A' is the set of attributes common to the objects in A, while B' is the set of objects which have all attributes in B. For example, for the context in Table 1.1, {4 6}′ = {a e g}, {c g}′ = {4 5}.

These two derivation operators can be combined in a pair of composite operators, denoted by $''$, which map $\wp(G)$ onto itself and $\wp(M)$ onto itself. For example, $\{4,6\}'' = \{4,6,7\}$, $\{c,g\}'' = \{c,e,g\}$.

If (G, M, I) is a context and A, A_1, A_2 are subsets of G, then

$$A_1 \subseteq A_2 \Rightarrow A_1' \supseteq A_2',$$
$$A \subseteq A'',$$
$$A'' = A'''.$$

Corresponding expressions hold for subsets of attributes of M. Therefore, the maps $' : \wp(G) \to \wp(M)$ and $' : \wp(M) \to \wp(G)$ induce a Galois connection between $\wp(G)$ and $\wp(M)$, when such sets are ordered with the set inclusion relation. It follows that the composite operators $''$ are closure operators.

A *concept* of the context (G, M, I) is a pair (A, B) where

$$A \subseteq G, B \subseteq M, A' = B, \text{ and } B' = A.$$

We call A the *extent* and B the *intent* of the concept (A, B). The set of all concepts of the context (G, M, I) is denoted by $\mathcal{C}(G, M, I)$. Thus, a concept is considered to be identified by its extent and its intent: the extent consists of all objects belonging to the concept while the intent contains all attributes shared by the objects. This formal definition of concept reflects one common meaning of concept in standard language with philosophical roots dating back to Aristotle. Essentially, it refers to the inverse relation between the number of attributes that are necessary to describe a concept and the number of objects to which the concept applies [217].

A consequence of the definition is that not all possible subsets of attributes (objects) are the extent (intent) of some concept. A subset A of G is an extent if and only if $A'' = A$; in this case the unique concept of which A is an extent is (A, A'). The corresponding statements applies to intents.

For instance, the subset {4 5} is a concept extent because $\{4\ 5\}'' = \{4\ 5\}$; the corresponding concept is (4 5, c e g). By contrast, (6, a e g) is not a concept of the context shown in Table 1.1; indeed there is no concept of which object 6 is the extent, because all the properties shared by object 6 are shared also by objects 4 and 7.

The intersection of any number of extents (intents) is always an extent (intent). In particular, it turns out that

$$\left(\bigcup_{j \in J} A_j\right)' = \bigcap_{j \in J} A_j',$$

$$\left(\bigcup_{j \in J} B_j\right)' = \bigcap_{j \in J} B_j'.$$

Conversely, the union of extents (intents) generally does not result in an extent (intent). For instance, take concepts (4 5 6 7, e g) and (1 2 5, e f) of the context in Table 1.1. The intersection of their extents is the extent of the concept (5, c e f g) and the intersection of their intents is the intent of the concept (1 2 3 4 5 6 7, e). The union of their extents or intents does not yield a concept's extent or intent.

The concepts of a context also permit a geometrical interpretation, because they can be seen as maximal rectangles of the table representing the context. More precisely, a *maximal rectangle* of the context (G, M, I) is a pair (A, B) where $A \subseteq G$, $B \subseteq M$, and such that

$$\forall g \in G \backslash A, \exists m \in M | (g, m) \notin I, \text{ and}$$

$$\forall m \in M \backslash B, \exists g \in G | (g, m) \notin I.$$

The last two conditions require that for each object not included in the maximal rectangle there exists at least one attribute of the maximal rectangle that is not shared by the object, and, correspondingly, that for each attribute not included in the maximal rectangle there exists at least one object of the maximal rectangle that does not share the attribute. For instance, (1 2, e f) and (3, d e h) are maximal rectangles (or concepts) of the context in Table 1.1, while (3, d e) violates the last condition.

An order relation on the set of concepts of a context is defined in the following way. If (A_1, B_1) and (A_2, B_2) are concepts in $\mathcal{C}(G, M, I)$, we say that (A_1, B_1) is a *subconcept* of (A_2, B_2), or that (A_2, B_2) is a *superconcept* of (A_1, B_1), and we write $(A_1, B_1) \leq (A_2, B_2)$, if $A_1 \subseteq A_2$ (which is equivalent to $B_1 \supseteq B_2$). We can also say that (A_1, B_1) is smaller (or more specific) than (A_2, B_2), or that (A_2, B_2) is larger (or more general) than (A_1, B_1). The relation $\leq$ is called the *hierarchical order* (or simply *order*) of the concepts.

The ordered set $\mathcal{C}(G, M, I; \leq)$ is called the *concept lattice* (or Galois *lattice*) of the context (G, M, I). It can be characterized by the following theorem.

The Basic Theorem on Concept Lattices. Let (G, M, I) be a context. Then $\mathcal{C}(G, M, I; \leq)$ is a complete lattice in which join and meet are given by:

$$\bigvee_{j \in J} (A_j, B_j) = \left(\left(\bigcup_{j \in J} A_j\right)'', \bigcap_{j \in J} B_j\right),$$

$$\bigwedge_{j \in J} (A_j, B_j) = \left(\bigcap_{j \in J} A_j, \left(\bigcup_{j \in J} B_j\right)''\right).$$

Conversely, if L is a complete lattice then L is isomorphic to $\mathcal{C}(G, M, I; \leq)$ if and only if there are mappings $\gamma : G \rightarrow L$ and $\mu : M \rightarrow L$ such that $\gamma(G)$ is join-dense in L, $\mu(M)$ is meet-dense in L, and gIm is equivalent to $\gamma(G) \leq \mu(M)$ for each $g \in G$ and $m \in M$. In particular, L is isomorphic to $\mathcal{C}(L, L, \leq)$ for every complete lattice L.

For a given complete lattice L, the context $(L, L, \leq)$ is not the only one whose concept lattice is isomorphic to L. In particular, consider the subsets $\mathcal{J}(L)$ of join-irreducible elements and $\mathcal{M}(L)$ of meet-irreducible elements. As they are join-dense and meet-dense in L, respectively, the Basic Theorem states that L is isomorphic to $\mathcal{C}(\mathcal{J}(L), \mathcal{M}(L), \leq)$, with the isomorphism given by

$$x \rightarrow (\downarrow x \cap \mathcal{J}(L), \uparrow x \cap \mathcal{M}(L)).$$

Turning to non-mathematical examples, in Figure 1.5 we show the concept lattice of the vertebrates context given in Table 1.1. The concept lattice contains as many as 16 concepts, including a top element with a non-empty intent (because all the animals have the 'has skeleton' property) and a bottom element with an empty extent (because there is no animal sharing all properties). Note that this is a very small fraction of the theoretical number of clusters that could be generated by considering all possible combinations of attributes and animals.

In fact, as a concept lattice contains only the pairs that are complete with respect to I according to the given definition, there are several subsets of properties or animals that cannot be admissible intents or extents for the given context. For instance, there cannot be any pair having an intent that

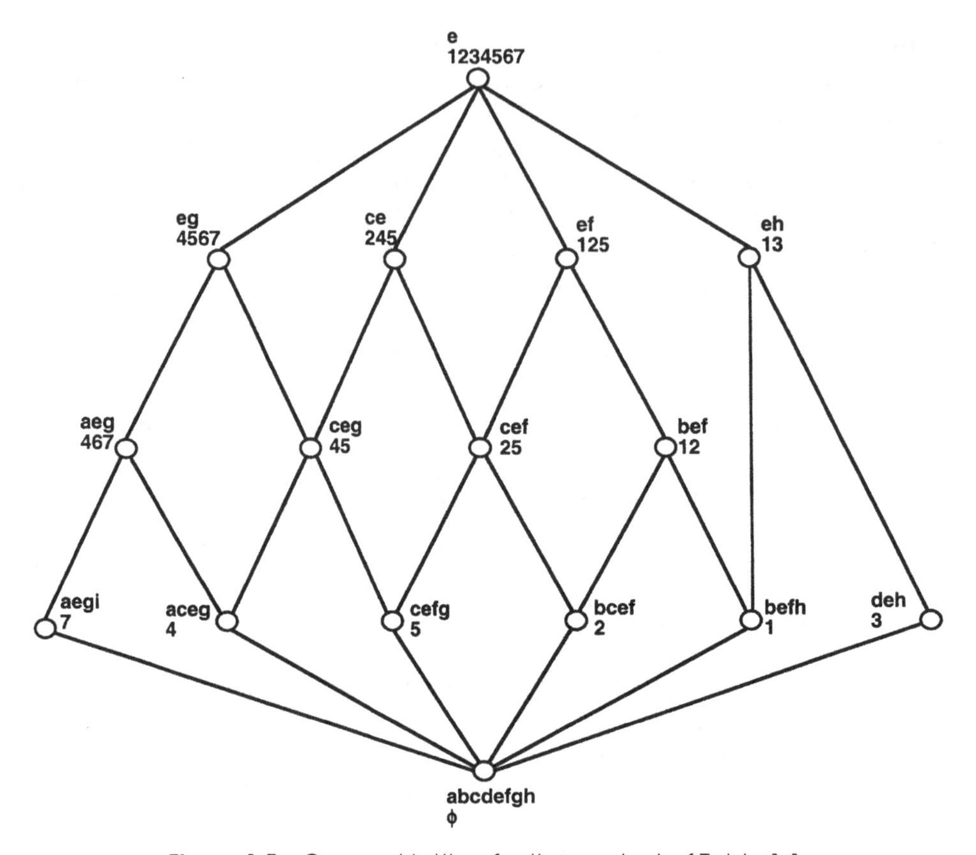

Figure 1.5 Concept lattice for the context of Table 1.1.

does not contain 'has skeleton', because all animals have 'has skeleton'; or, to take another example, there cannot be any pair having an intent equal to 'can fly' and 'has skeleton' because any animal that 'can fly' also 'has skeleton' and 'has wings'.

On the other hand, the similarities and the diversities of the animals described in the context are well represented in the concept lattice. On closer inspection, some concepts in Figure 1.5 correspond to well-known vertebrate groups such as bird (a group which 'has wings' and 'has beak'), fish (a group which 'breathes in water'), and mammal (which 'is viviparous', among other things). Other concepts arise from sharing a more restricted set of attributes (e.g., being able to fly without necessarily being a bird, or living in water without necessarily breathing in water) or from mixing attributes belonging to different classical groups (e.g., having a beak and living in water).

This situation reflects quite well the problems encountered in the study of zoological taxonomy (i.e., the classification and naming of animals), where different choices of the attributes used for classifying animals may lead to different classifications over the same set of animal data. In a concept lattice, it is as if many possible hierarchical classifications (composed of overlapping data sets) existed in parallel with each other with no duplication of information (as opposed to decision trees, for instance, where the same subtree may occur multiple times over the whole structure). The concept lattice thus represents a formalism for exploring such hierarchies for correlations, similarities, refinements, anomalies, or even inconsistencies. We will see many applications of this general exploratory paradigm throughout this book.

An important issue for applications is how to label the set of concepts. Figure 1.5, where the intent and the extent of every concept are shown, contains much redundant information because, for each concept (X, Y), X appears in every ancestor of (X, Y) and Y appears in every descendant. This may be useful, but it is not strictly necessary.

A more compact form of labelling can be obtained by making use of object concepts and attribute concepts. The *object concept* of an object $g \in G$ is the concept (g'', g'), where g' is the *object intent* $\{m \in M | gIm\}$ of g. The object concept of g, denoted $\gamma(g)$, is thus the smallest concept with g in its extent. Correspondingly, (m'', m') is the *attribute concept* of an attribute $m \in M$, m' being the *attribute extent* $\{g \in G | gIm\}$ of m. The attribute concept of m, denoted $\mu(m)$, is the largest concept with m in its intent.

The object concept of g and the attribute concept of m can also be characterized by the following expressions:

$$\gamma(g) = \bigwedge \{(X, Y) \in \mathcal{C}(G, M, I) | g \in X\},$$

$$\mu(m) = \bigvee \{(X, Y) \in \mathcal{C}(G, M, I) | m \in Y\}.$$

By using object concepts and attribute concepts, each object and each attribute of the context can be displayed in the line diagram only once, namely only for the respective object or attribute concept. Still, the extents and the intents of the other concepts can be read off from the diagram with reduced labelling. The extent of each concept is formed by collecting all objects which can be reached by descending line paths from the concept. The intent of each concept consists of all attributes located along ascending line paths. As an illustration, Figure 1.6 shows the concept lattice of the context in Table 1.1 with reduced labelling.

One advantage of such a representation is that we can enter the full names of objects and attributes into the diagram line, thus improving its readability (see Figure 1.6). Concept lattices with reduced labelling are also termed inheritance concept lattices in the software engineering field, where they have been used to analyse class interfaces (e.g., [64], [101]).

Just as a set of concepts can be uniquely determined from a given context, so too can the context be reconstructed from its concepts. The set G is the extent of the largest concept (G, G'), the set M is the intent of the smallest

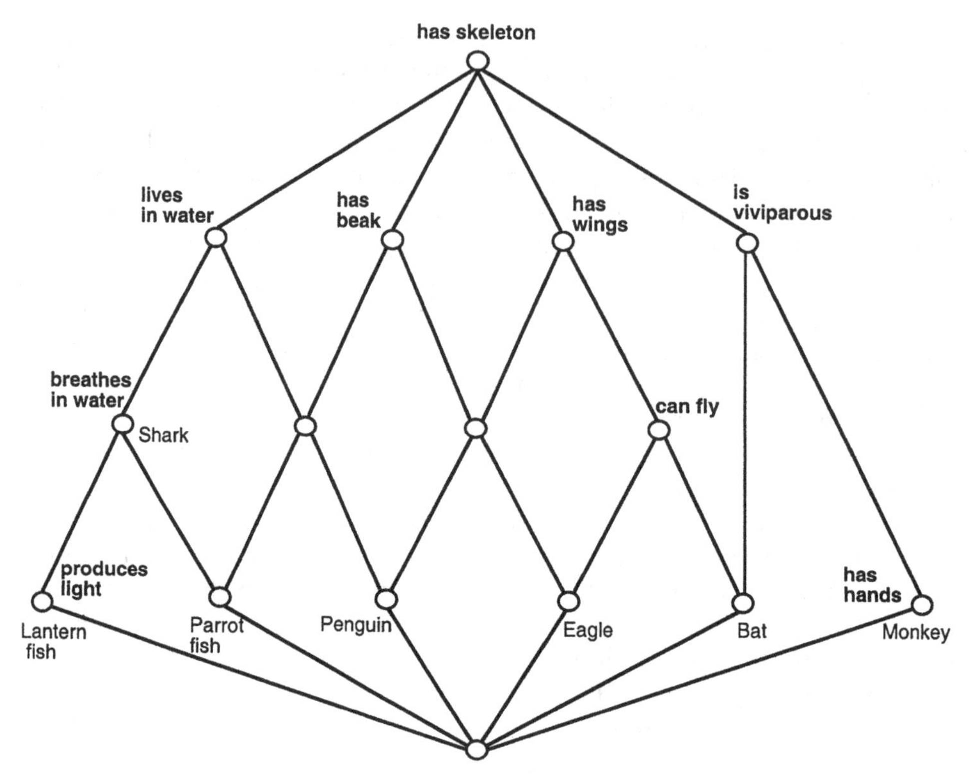

Figure 1.6 Concept lattice for the context of Table 1.1 with reduced labelling.

concept (M', M), and the incidence relation I is given by

$$I = \bigcup \{A \times B \mid (A, B) \in \mathcal{C}(G, M, I)\}.$$

Reading off the context from the full concept lattice (i.e., from the set of concepts plus the line diagram) is even easier. As the Basic Theorem shows:

$$(g, m) \in I \iff \gamma(g) \leq \mu(m).$$

Different contexts can yield isomorphic concept lattices. One simple context manipulation which does not alter the structure of the concept lattice consists of merging the objects with the same intents and the attributes with the same extents, respectively. This operation is usually referred to as *clarification*.

Another structure-preserving transformation is the removal of *reducible objects*, i.e., objects whose intent can be written as the intersection of other object intents, and *reducible attributes*, i.e., attributes whose extent can be written as the intersection of other attribute extents.

It turns out that in a finite concept lattice, it is possible to omit all reducible objects and attributes, because, as seen in Section 1.1, each element is the join of join-irreducible and the meet of meet-irreducible elements. In a sense, irreducible elements can be seen as the building blocks of a concept lattice, although this characterization, termed the *reduced* concept lattice, has mainly theoretical interest.

Before concluding this section, we would like to remark that the duality principle naturally extends to concept lattices. In particular, if we exchange the roles of objects and attributes, we obtain the dual concept lattice.

1.3 Many-valued contexts

Many objects in the real world are naturally described by attributes which can take on several values, such as 'shape' or 'colour' or 'birth year', rather than by the presence or the absence of some properties. The former attributes are called *many-valued attributes*, in contrast to the *one-valued attributes* considered so far.

A *many-valued context* is a quadruple (G, M, V, I) consisting of a set G of objects, a set M of many-valued attributes, a set V of attribute values, and a ternary relation I between G, M, and V (i.e., $I \subseteq G \times M \times V$) such that

$$(g, m, v) \in I \text{ and } (g, m, w) \in I \text{ always implies } v = w.$$

The expression $(g, m, v) \in I$ reads 'the attribute m has the value v for the object g'; alternatively, as the many-valued attributes can be regarded as partial maps from G into V, we can write $m(g) = v$.

Note that a many-valued context is like a database relation [152]; the presence of both the attributes and the attributes values is explicitly taken into account to facilitate the extension of the concepts seen in the previous section to objects described by many-valued attributes.

As an illustration of a many-valued context, we will give an example taken from [69]. Table 1.2 shows a simple many-valued context describing the planets of our solar system. The nine planets known at the time (1990) are described by three many-valued attributes, namely 'size' (small/medium/large), 'distance from the sun' (near/far), and 'moon' (yes/no).

In order to assign concepts to a many-valued context, we can transform the many-valued context into a single-valued one, and then interpret the concepts of the derived context as the concepts of the many-valued context.

One simple way to do such a transformation is to replace every many-valued attribute by the corresponding attribute–value pairs, with each object being described by one attribute–value pair per many-valued attribute. In our example, for instance, the planet Mercury can be seen as described by

Table 1.2 A context for the planets

	size			*distance from sun*		*moon*	
	small	medium	large	near	far	yes	no
Mercury	×			×			×
Venus	×			×			×
Earth	×			×		×	
Mars	×			×		×	
Jupiter			×		×	×	
Saturn			×		×	×	
Uranus		×			×	×	
Neptune		×			×	×	
Pluto	×				×	×	

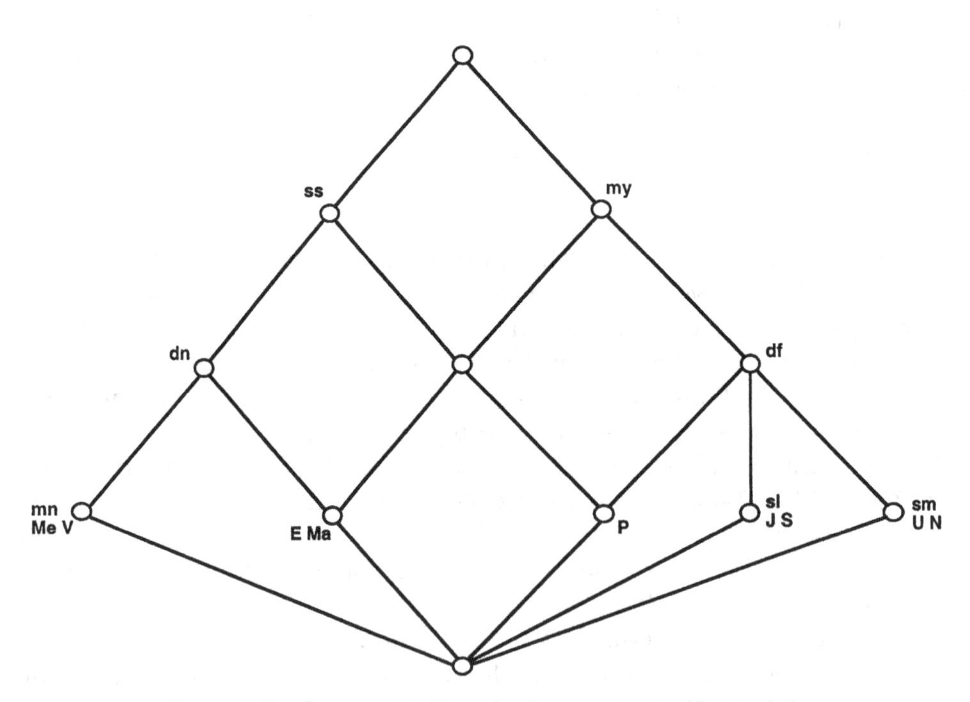

Figure 1.7 Concept lattice for the context of Table 1.2.

the three single-valued attributes: 'size–small' (*ss*), 'distance–near' (*dn*), and 'moon–no' (*mn*). If we think of a bit-vector representation, each attribute is assigned a set of bits (one for each value of the attribute), no more than one of which will be on.

The concept lattice of the context in Table 1.2 is shown in Figure 1.7. In this case the top element has an empty intent, because there is no attribute–value pair shared by all objects, and there are different objects which give rise to the same object concepts, i.e., those objects with the same object intent.

Note that all object concepts are parents of the bottom element, in contrast to the concept lattice for the single-valued context seen earlier, where some object concept was a superconcept of another object concept (see Figure 1.5). This feature is a consequence of the particular transformation being used, characterized by the fact that each object is described by all many-valued attributes and that the values of each attribute mutually exclude each other. For the same reasons, the bottom element will always have an empty extent.

Considering mutually exclusive attribute values works well in many circumstances but it may be unsatisfactory in certain cases, especially when the values taken on by each attribute are ordered, as for instance with numerical attributes. Consider again the planets context, but this time assume that the attribute 'distance from sun', expressed in millions of kilometres, is given by:

$$
\begin{aligned}
d(\text{Mercury}) &= \textit{tens},\\
d(\text{Venus}) &= \textit{few hundreds},\\
d(\text{Earth}) &= \textit{few hundreds},\\
d(\text{Mars}) &= \textit{few hundreds},\\
d(\text{Jupiter}) &= \textit{several hundreds},\\
d(\text{Saturn}) &= \textit{thousands},\\
d(\text{Uranus}) &= \textit{thousands},\\
d(\text{Neptune}) &= \textit{thousands},\\
d(\text{Pluto}) &= \textit{thousands}.
\end{aligned}
$$

This description is more detailed but consistent with that provided for the same attribute in Table 1.2, in the sense that planets which were formerly distinct according to their distance from the sun will remain distinct.

For the new many-valued context, it may be convenient to derive the single-valued context in a different way. In particular, we would like to express the fact that some values subsume other values because the former are greater (or smaller) than the latter. This can be done by creating new attribute values with the required property from the original values. A derived single-valued subcontext which conceptually models the greater-than relationship between the values of the 'distance from sun' attribute is shown in Table 1.3.

The concept lattice of the full derived planet context including the subcontext of Table 1.3 is shown in Figure 1.8. In this case, the greater-than

Table 1.3 A derived subcontext for the attribute 'distance from sun' with numerical values

	distance from sun (millions of kms)			
	≥ tens	≥ few hundreds	≥ sev hundreds	≥ thousands
Mercury	×			
Venus	×	×		
Earth	×	×		
Mars	×	×		
Jupiter	×	×	×	
Saturn	×	×	×	×
Uranus	×	×	×	×
Neptune	×	×	×	×
Pluto	×	×	×	×

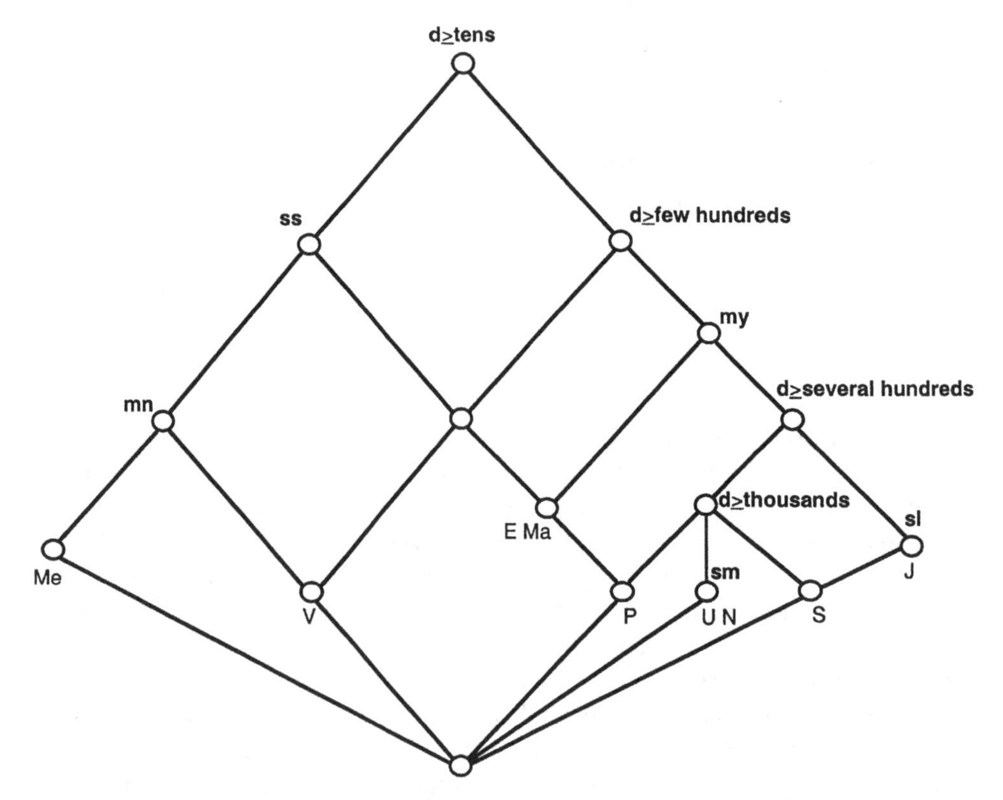

Figure 1.8 Concept lattice for the context of Table 1.3.

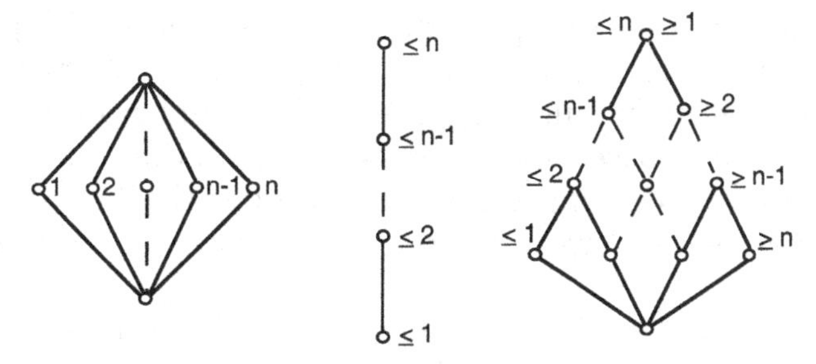

Figure 1.9 The concept lattices of a nominal, ordinal and interordinal scale.

relationship between the values taken on by the 'distance from sun' attribute has been maintained in the full concept lattice, where the concepts described by those attribute values have been ordered accordingly.

Thus, for instance, one can easily find the planets whose distance from the sun is greater than or equal to a few hundreds of millions of kilometers as a subconcept of the planets whose distance is greater than or equal to tens of millions.

Other kinds of many-valued attributes may give rise to different derivations of the corresponding single-valued context. In Section 2.4.1, for instance, we will see an example of a context with tree-structured attributes. More generally, for each many-valued attribute m one can provide a table which specifies how to transform every value of m into one or more new values. This process is called *conceptual scaling* by Ganter and Wille [96], because such a table can be seen as a particular context (or conceptual scale), with scale values and scale attributes.

The scales corresponding to the two interpretations of the 'distance from sun' attribute seen above are called nominal and ordinal, respectively. The concept lattices for a nominal, ordinal and interordinal (i.e., with an interval structure) n-element scale are shown in Figure 1.9.

1.4 Bibliographic notes

There are many textbooks dealing with the mathematical theory of lattices and ordered sets, of which the notions presented here in Section 1.1 represent a small fragment. Well-known representative texts are [24], [68] and [69]. The theoretical foundations of concept lattices as a tool for data analysis are more recent. Although the first breakthrough was made in 1940 by Birkhoff, who proved that a lattice can be constructed from every binary relation between some objects and some attributes [24], it was only more recently that the potential applications of Birkhoff's results were made more explicit [16] and then further elaborated into more comprehensive studies, especially by Rudolf Wille and his co-workers in Darmstadt.

The Basic Theorem on Concept Lattices is presented in [253], conceptual scaling in [95] and [261], implications and dependencies between attributes in [93] and [259]. These findings, together with a number of more advanced mathematical features based on order theory, including parts and factors [256], composition and decomposition operators ([254], [257], [258], [92]), and structural properties of concept lattices, were summarized by Ganter and Wille [96].

The material presented here is mainly based on the first three chapters of [96] and on Chapters 1, 2, and 11 in [69]. Readers interested in more sophisticated aspects of the mathematical theory of lattices and concept lattices should consult the references cited above.

Some researchers have recently attempted to generalize the basic model of concept lattices to deal with objects which are not simply described by a set of attributes, whether single- or many-valued. It is conceivable to use more powerful representation formalisms that would allow us, for instance, to assign weighted attributes to objects, or to express negation or disjunction, or to use structured objects formed by parts and relationships over parts.

Works in this research line include triadic concept analysis ([142], [262]), where a ternary incidence relation is used that combines objects, attributes, and conditions under which objects may have certain attributes, the extension of concept lattices to objects described by non-atomic attributes ([146], [77]), and logical concept analysis ([51], [80], [52]). The last approach was inspired by the computational logic paradigm; objects and concepts are described by logical formulae and thus the subsumption relation between intents can no longer be expressed in set-theoretical terms but requires some form of logical implication. These exploratory attempts, however, confront both theoretical and computational difficulties and are presently beyond the reach of practical software implementation.

Also explored are the relationships and synergies with other knowledge representation formalisms such as description logics and conceptual graphs ([12], [263]), including the use of concept lattices to support ontology engineering [227].

Finally, it is worth noting that forming hierarchies of concepts with intensional descriptions has been studied in several domains (e.g., [165], [97], [156], [85], [131]). In the machine learning and data mining fields, this is usually referred to as conceptual clustering [82], after Michalski's pioneering work [162]. Most conceptual clustering approaches do not rely on a clear theoretical background; equipped with a set of hierarchy-change operators, they typically perform a hill-climbing search through the space of possible concept hierarchies to find a 'good' hierarchy. As the criteria they use are local and/or heuristic, these methods are prone to errors and are, in general, sensitive to the order in which the objects are considered. On the other hand, they may employ a more expressive or flexible representation language than the strictly conjunctive logical expressions used in a concept lattice, such as probabilistic descriptions ([81], [156]), contingency tables [117], predicate calculus [221], or first-order logic [25].

2

Algorithms

In this chapter we present a set of algorithms for solving specific tasks occurring in the development of applications based on concept lattices. The tasks covered include the construction and maintenance of concept lattices, their visualization on a computer screen, and their transformation to account for additional existing knowledge.

We have not attempted to cover all the literature available on this subject. Various algorithms have been proposed for each main task addressed in the chapter, and most of these algorithms have then been progressively improved or extended.

We focus on some fundamental strategies that may have been realized or popularized in different guises in the literature. The description of each algorithm provided here may closely reflect some earlier proposal, or it may have been obtained by assembling parts of different earlier algorithms or by making them more precise. In either case, the presentation is intended to be detailed enough to be easily implemented and to ease cross-comparison. In the bibliographic notes at the end of the chapter, the papers which first reported the original version of each algorithm described, if any, are referenced. Such papers usually also contain a formal proof of the correctness of the algorithm, which is not reported here.

In the illustration of the algorithms, the following rules and notation will be used. The description is given in pseudo-code, indentation is used to denote the beginning and the end of the statements, and the control statements are printed in bold. When their implementation is straightforward, certain functions may be expressed in natural language. The statements are then enumerated for reference in the main text. Variables and functions can be distinguished by the first letter of their name: in lower and upper case, respectively.

Concept Data Analysis: Theory and Applications. Claudio Carpineto and Giovanni Romano
 ISBN: 0-470-85055-8

2.1 Constructing concept lattices

In this section we will address the question of the efficient construction of concept lattices. After analysing the size of the lattice as a function of the input size, we will present a number of algorithms catering for the most common situations that can be encountered. More specifically, we will consider three main tasks: (i) determination of the whole set of concepts; (ii) batch construction of the full concept lattice including the Hasse diagram; and (iii) generation of partial concept lattices for tasks where it is sufficient to generate a limited portion of the whole structure.

2.1.1 Computational space complexity of concept lattices

There are some theoretical upper bounds on the number of concepts present in the lattice that depend on the size of the input context. If a context is described by $|G|$ objects and $|M|$ attributes, the corresponding lattice will contain at most $2^{|G|}$ or $2^{|M|}$ concepts, whichever is the largest. For many-valued contexts, assuming that each object is described by $|M|$ attributes and v values per attribute, the latter bound becomes $(v+1)^{|M|}$. A better upper bound can be expressed in terms of the number of incident relations $|I|$; for $|I| > 2$, there may be at most $\frac{3}{2} \cdot 2^{\sqrt{|I|+1}} - 1$ concepts [210].

It turns out that such high theoretical bounds can be reached for some actual contexts. A well-known example is the $(n \times n)$-dimensional context containing ones in all positions, but zeros along the diagonal; the corresponding lattice contains exactly 2^n concepts [176].

However, this situation may occur on rare occasions (in fact, it is an indication that the context has been ill designed). It is also often the case that the number of attributes per object has upper bound given by a constant K. In this case, since each new object can generate at most 2^K concepts (i.e., all possible subsets of its intent), the number of concepts is bounded by $|G| \cdot 2^K$. With this assumption, the growth of the number of concepts is therefore linear with the number of objects, but the factor may become very large even for relatively small values of K. One can hypothesize that the ratio between the number of concepts $|C|$ and $|G|$ is usually much smaller than this upper bound and more stable with respect to attribute variation.

To gain some deeper insights into the order of magnitude of the lattices, it is convenient to study this problem for the case when the object description obeys some simple distribution of probability. We hypothesize that each attribute is assigned to each object with constant, independent probability $p = k/|M|$. Under this assumption, the number of keywords per object follows a binomial distribution with a mean value of k.

The mean number of concepts can be computed in the following way. We need to sum the probabilities of occurrence of all possible concepts, which can be conveniently partitioned into subsets of concepts characterized by a fixed number of objects and of attributes. There are $\binom{|G|}{i}\binom{|M|}{j}$ possible concepts with i objects and j attributes, where each concept has the same probability of occurrence. This probability can be obtained as the product of three factors, namely the probability that the i objects share the j attribute values (i.e., p^{ij}), the probability that the i objects do not share other attributes (i.e., $(1-p^i)^{|M|-j}$), and the probability that there exists no other object having the j attributes (i.e., $(1-p^j)^{|G|-i}$). The complete formula is given by

$$|C| = \sum_{i=0}^{|G|} \sum_{j=0}^{|M|} \binom{|G|}{i} \binom{|M|}{j} p^{ij}(1-p^i)^{|M|-j}(1-p^j)^{|G|-i} \tag{2.1}$$

Figure 2.1 shows results plotted from equation (2.1) for four values of k (5, 10, 20 and 50), choosing $|G| = 10\,000$. The curves suggest that the number of concepts varies from linear to quadratic with respect to the number of objects, at least for the chosen range of values for k. The size of the lattice grows as the number of attributes per object increases; the upper bound is reached for

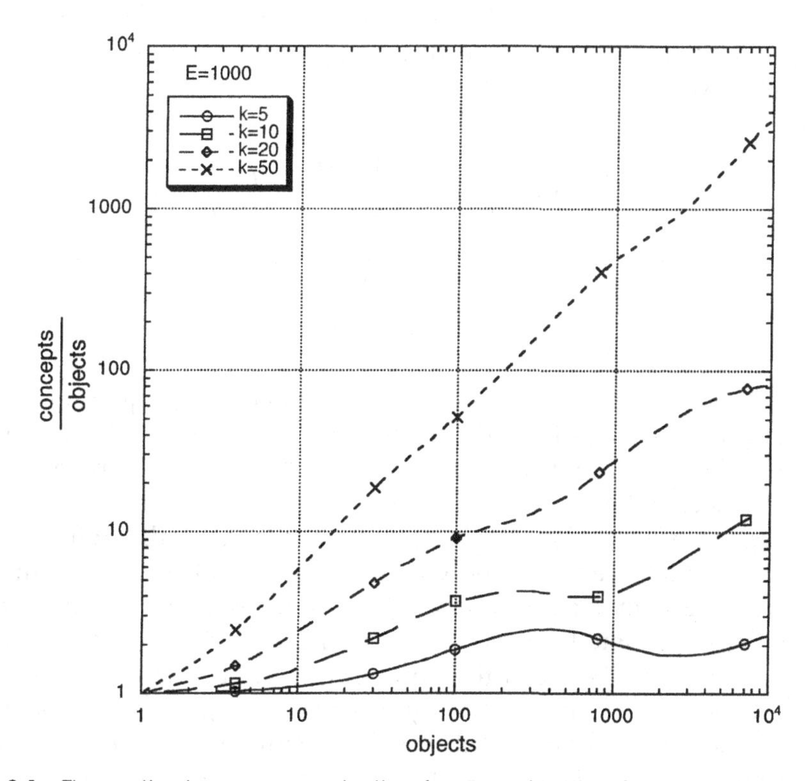

Figure 2.1 Theoretical space complexity of concept lattice for one-valued context. Both scales are logarithmic.

$k = 50$, corresponding to a probability of assigning an attribute to an object equal to $50/1000 = 0.05$. The latter value accounts for a relatively dense context table, at least for information retrieval applications, which represents a realistic application domain where objects are documents and attributes are keywords.

A formula can be derived also for multi-valued contexts, assuming that each object is described by a fixed set of $|M|$ attributes with v values each, to investigate the effects that the two parameters $|M|$ and v (governing the size of the descriptor space) have on the size of the lattice.

Under the assumption of a uniform distribution of the attribute values, a formula analogous to equation (2.1) can be found. The main differences are that the probability that one object has a certain attribute value is equal to $1/v$, instead of to $k/|M|$, and that the number of concepts with i objects and j attributes is given by $\binom{|G|}{i}\binom{|M|}{j}v^j$.

A further minor modification is needed to deal with the bottom element of the lattice. Its contribution cannot be expressed as an addendum of the general summation (i.e., for $i = 0$, $j = |M|$, in equation (2.1)), because this would produce an indeterminate form. Thus, we need to restrict the range of the summation index and add one external addendum.

For the sake of symmetry, we treat also the contributions of the top element of the lattice separately. In the following equation, the last two terms account for the probabilities of occurrence of the top and bottom elements of the lattice, respectively:

$$|C| = \sum_{i=1}^{|G|}\sum_{j=1}^{|M|}\binom{|G|}{i}\binom{|M|}{j}v^j\left(\frac{1}{v}\right)^{ij}\left(1-\left(\frac{1}{v}\right)^{i-1}\right)^{|M|-j}$$

$$\times\left(1-\left(\frac{1}{v}\right)^{j}\right)^{|G|-i} + \left(1-\frac{1}{v^{|G|-1}}\right)^{|M|} + \left(1-\frac{1}{v^{|M|(|G|-1)}}\right) \qquad (2.2)$$

Figure 2.2, taken from [40], shows results plotted from equation (2.2) for four pairs of values for $(|M|, v)$: (5, 5), (10, 10), (50, 10), (50, 50). Even in this case, the number of concepts grows at most quadratically with respect to the number of objects. The space complexity moves towards the upper bound when the ratio between M and V increases, or, less markedly, when both M and V grow by the same factor and the number of objects is not too small.

The lower curve in Figure 2.2 shows what happens when the concept description space is small relative to the number of objects. After reaching a maximum, the ratio between the number of concepts and the number of objects decreases. In the limit, as $|G|$ grows to infinity, $|C|/|G|$ tends to zero because, as said earlier, for $|M|$ attributes and v values per attribute, $|C|$ has an upper bound of $(v+1)^{|M|}$. In this case, the theoretical upper bound is $6^5 = 7776$; thus, after a certain point the lattice is saturated and $|C|/|G|$ drops below one.

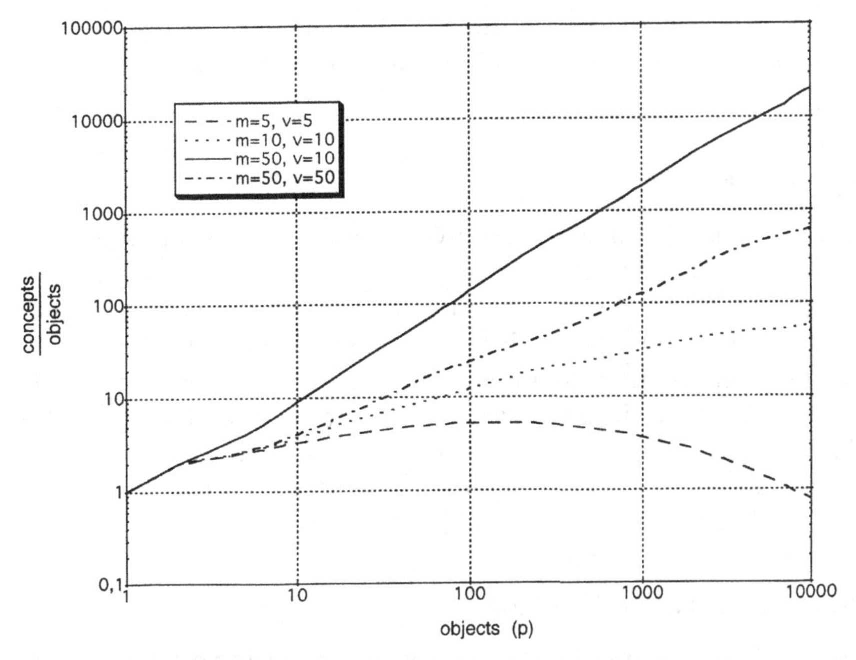

Figure 2.2 Theoretical space complexity of concept lattice for many-valued context. Both scales are logarithmic.

It is important to note that these results hold when the attributes are assigned to the objects with independent probability. This is a rather strong assumption, which may not hold in many cases (e.g., for contexts obtained by ordinal scaling). On the other hand, these theoretical findings have been confirmed by the results of a number of experiments. There is a large body of evidence (e.g., [40], [214], [102], [46], [44]) that shows that the time complexity of lattice construction grows polynomially – usually quadratically – with respect to the number of objects, whereas exponential behaviour is very rare. In Part II of the book we will see several examples concerning one-valued document collections and many-valued natural data sets.

In the following sections, we will present several algorithms for generating all concepts of a given (single-valued) context, with or without the line diagram. As the output size can be exponential in the size of the input context, we will characterize their asymptotic complexity as a function of the number of concepts generated. For the algorithms described below, theoretical bounds of this kind will be derived and used for comparison.

2.1.2 Construction of the set of concepts

In this section we describe three basic algorithms for finding the set of concepts associated with a given context. Although they may not be the best available algorithms from an application point of view, they very clearly illustrate some fundamental computational problems presented by the task at hand. In addition, they may be used as the building blocks for more powerful algorithms.

The Naive algorithm

As seen in Section 1.2, a subset A of G is the extent of some concept if and only if $A'' = A$, in which case the unique concept of which A is an extent is (A, A'). Dually, a subset B of M is the intent of some concept if and only if $B'' = B$, the unique concept of which B is an intent being (B', B).

Thus, the simplest algorithm for generating the set of concepts of a formal context (G, M, I) would be to form (A'', A') for all $A \subseteq G$, or (B', B'') for all $B \subseteq M$. This method is extremely inefficient as it requires the consideration of all the subsets of G or M together with their corresponding closures, regardless of the size of the concept lattice being generated.

Ordering subsets based on their closure

A better strategy for finding the set of concepts by using closures is based on ordering the subsets of G in such a way that only some subsets need to be examined.

After imposing a linear order, denoted by '$<_G$', on the elements of G, all subsets of G may, in turn, be strictly ordered according to a particular lexicographic order ($\preceq$) defined as follows. A subset A of G is smaller (according to $\preceq$) than a subset B of G if the smallest element which distinguishes A and B according to $<_G$ belongs to B. Note that this is an order on bit strings, not on natural numbers. For instance, $\{4\ 7\} \preceq \{4\ 6\}$ because the smallest distinct element (i.e., 6) belongs to $\{4\ 6\}$. To illustrate, the first 16 extents for the context in Figure 1.1 according to $\preceq$ are shown in Figure 2.4.

From a computational point of view, the lexicographic order can be generated in the following way. The next subset is generated from the current subset in two steps, i.e., by adding the maximum object of G that is not contained in the current subset and then deleting all objects in the current subset that are greater than the object just added.

This particular enumeration of subsets of G is well formed with respect to the properties of closures. It turns out that, for any subset of objects A, if $A''\backslash A$ contains only elements that are greater ($<_G$) than those in A, then all subsets that follow A and precede A'' in the lexicographic order $\preceq$ will produce a concept extent that is either equal to A'' or will be generated later by some subset following A''. Thus, in this case it is possible to skip examination of some of the subsets of G. If the condition about the difference between A'' and A is not satisfied, then the creation of the concept with extent A'' will be postponed until the subset A'' is encountered, thus avoiding multiple generation of the same concept.

The algorithm considers one current subset A in the lexicographic order at a time, until the last element of the order (i.e., the subset G) is generated. If A'' does not contain objects that are smaller than those in A, which is trivially true if $A'' = A$, then the concept with extent A'' is added to the set of concepts and the next subset of G to be examined is set to the successor of A''. If the condition is not satisfied, the concept with extent A'' is not added to the list of concepts because it will at the latest be generated when examining the subset A'', and the next subset in the lexicographic order is generated.

The algorithm is shown in Figure 2.3. To illustrate the Next Closure algorithm, as well as the other algorithms which will be presented below, we will use the context in Table 1.1. In Figure 2.4, we show the list of concept extents generated while enumerating the first 16 subsets of objects in lexicographic order.

In the example given, the algorithm examines 14 of the 16 subsets and generates 7 concepts. The number of closures generated will therefore usually

Next Closure
Input: Context (G, M, I)
Output: The set C of all concepts of (G, M, I)

1. $C := \{(M', M)\}$
2. $currSubset := \max\{g \in G\}$
3. $nextObj := \max\{g \in G\}$
4. **while** $currSubset \neq G$
5.   **if** there is no $g \in currSubset'' \setminus currSubset$ such that $g < nextObj$
6.   **then**
7. $C := C \cup \{currSubset'', currSubset'\}$
8. $nextObj := \max\{g \in G \setminus currSubset''\}$
9. $currSubset := currSubset''$
10.   **else**
11. $nextObj := \max\{g \in G \setminus currSubset$ such that $g < \max(currSubset)\}$
12. $currSubset := currSubset \cup \{nextObj\}$
13. $currSubset := currSubset \setminus \{g \in currSubset$ such that $nextObj < g\}$

Figure 2.3 The Next Closure algorithm for finding the set of concepts.

7 6 67 5 57 56 5674 47 46 467 45 457 456 4567 3 ...
↓ ↓ ↓ ↓ ↓ ↓ ↓ ↓ ↓ ↓ ↓ ↓ ↓ ↓
(7) () () (5) () () () (4) () (467) (45) () (4567) (3)

Figure 2.4 Step-by-step computation of the first concept extents for the context in Table 1.1 by the Next Closure algorithm. A '()' denotes the calculation of a closure that has not been added to the set of concepts.

be greater than the number of concepts and smaller than the number of subsets, although in the worst case the algorithm may have to explore all subsets of G. An example is the context where each object has the same description; the concept lattice consists of the single element (G, M) and the Next Closure algorithm must generate all the subsets preceding G. The worst-case time complexity of the algorithm is thus $O(2^{|G|}|G||M|)$, where $O(|G||M|)$ is the time required to compute the prime ($'$) and doubleprime ($''$) operators.

One advantage of this algorithm is that every concept is created only once, and thus no extra space and time is needed to store and retrieve the generated concepts. A main limitation, besides its relative efficiency, is that the algorithm does not support the construction of the Hasse diagram, for there is no relationship between the lexicographic order and the order defined over the concepts in the lattice.

Computing intersections incrementally

A different solution to the problem of finding all the concepts is based on the observation that every concept extent is the intersection of attribute extents and every concept intent is the intersection of object intents.

To generate the set of concepts, it is therefore sufficient to form all possible intersections between the sets of objects associated with each attribute and then use the context to find the intent corresponding to each generated concept extent. Alternatively, we could calculate all intersections of object intents and then determine the extent of each such concept intent.

In order to form all possible intersections, it is convenient to consider one attribute (object) at a time and calculate the intersection between its extent (intent) and each concept extent (intent) in the current set of concepts, where the current set of concepts contains concepts generated by the attributes (objects) that have already been examined.

When dual versions of the same algorithms are possible, as in this case, the choice should be driven by the consideration of how the size of G and M affects algorithm efficiency. We will see other examples of this in the following.

The pseudo-code for the determination of all concepts is shown in Figure 2.5. This version iterates over the set of objects; it is termed *Object Intersections*.

The working of the algorithm is illustrated using again the animal context in Table 1.1. The set of new concepts generated for each attribute of the context is shown in Figure 2.6. Note that in Figure 2.6 extents and intents are represented without brackets, to keep the notation simple; this notation will also be used in the rest of the book provided that the meaning is clear.

Object Intersections
Input: Context (G, M, I)
Output: The set C of all concepts of (G, M, I)

1. $C := \{(M', M)\}$;
2. **for** each $g \in G$
3. **for** each $(X, Y) \in C$
4. $Inters := Y \cap \{g\}'$;
5. **if** *Inters* is different from any concept intent in C **then**
6. $C := C \cup \{(Inters', Inters)\}$

Figure 2.5 The Object Intersections algorithm for finding the set of concepts.

M: (∅, a b c d e f g h)
1: (1, b e f h)
2: (2, b c e f), (1 2, b e f)
3: (3, d e h), (1 3, e h), (1 2 3 4 5 6 7, e)
4: (4, a c e g), (2 4 5, c e)
5: (5, c e f g), (1 2 5, e f), (2 5, c e f}, (4 5, c e g)
6: (6, a e g)
7: (7, a e g i)

Figure 2.6 List of concepts generated for each object by the Object Intersections algorithm for the context in Table 1.1.

As the Object Intersections algorithm processes one object at a time, it is also suitable for incrementally determining a concept lattice, as will be better explained in Section 2.2.1.

The major drawbacks of the Object Intersections algorithm are that the same concepts can be generated several times and no provision is made for generating the concept lattice. The worst-case time complexity is as follows. Line 4 can be done in $O(|M|)$. Line 5 would take $O(|M||C|)$ time, but its cost can be significantly reduced by maintaining a search tree or other data structure for set manipulation problems [5]. One very useful search structure, which will be used throughout the book, is a *trie*. A trie, pronounced 'try', is essentially an N-ary tree whose nodes are N-place vectors with components corresponding to digits or characters. Each node on level l represents the set of all keys that begin with a certain sequence of l characters and specifies an N-way branch, depending on the $(l + 1)$th character [134]. The name 'trie' was suggested because it is a part of information re*trie*val. Such an external structure, of course, adds to the memory requirements, but it allows us to store and retrieve concepts in $O(|M|)$ time, using the intent of the concept as the key. As line 4 and line 5 are repeated at most $|G||C|$ times, their contribution to the final complexity is $O(|G||C||M|)$. Line 6 can be computed in $O(|G||M|)$, and since this operation is performed exactly $|C|$ times its contribution is $O(|G||C||M|)$. Thus, the overall complexity is $O(|G||C||M|)$.

2.1.3 Construction of concept lattices

The algorithms introduced in the previous section are able to determine the set of concepts of a given context but they are unable to find the line diagram at the same time. Now we address the problem of building the full concept lattice, which is a prerequisite for many applications.

Finding neighbours

One of the best-known and conceptually simplest algorithms to construct the set of concepts along with the Hasse diagram is based on generating neighbours iteratively, according to the $\prec$ relation.

Starting from the top element of the lattice (G, G'), the algorithm builds one level at a time, where the next level contains the children of all concepts present in the current level. More specifically, for each concept in the current level, a function (described below) is invoked that calculates the lower neighbours of that concept; then it is checked if each returned child has not been already generated, in which case the concept is added to the lattice, and the concept is finally linked to its parent.

The sequence of concepts generated this way corresponds to a top-down breadth-first visit of the final lattice, although a depth-first implementation would work equally well. The pseudo-code of the algorithm, called *Next Neighbours*, is shown in Figure 2.7. For the sake of generality, the concept lattice is seen as a set of concepts C and of a set of edges E, where the edges are ordered pairs of concepts (c_1, c_2) such that $c_1 \prec c_2$, i.e., c_1 is a lower neighbour of c_2. In practice, however, it is convenient to implement each concept as a record with pointers to its neighbours.

The calculation of the children of a given node is performed by the function *FindLowerNeighbours*. The implementation of this function is based on the observation that all lower neighbours of a concept (X, Y) are contained in a small set of concepts, each of which is formed by adding just one new attribute m to Y ($Y_1 = Y \cup \{m\}$) and then by computing (Y_1', Y_1''). At this point, by comparing the candidates generated, one could select only those with the largest Y_1', thus obtaining the actual set of lower neighbours.

If an auxiliary search tree is used to check whether a concept has already been generated (line 9 in Figure 2.7), the worst-case time complexity of the algorithm is obtained by multiplying the number of invocations of the function *FindLowerNeighbours* (which is equal to the total number of concepts) by the cost of computing the lower neighbours of a given node. The latter requires the calculation of the prime operator for generating the candidates (line 3 and 4 in function *FindLowerNeighbours* in Figure 2.7), which takes $O(|G||M|)$ time, for at most $|M|$ times. As the final removal of non-maximal candidates takes at most $O(|G||M|^2)$ time, the overall complexity of the

NextNeighbours
Input: Context (G, M, I)
Output: The concept lattice $L = (C, E)$ of (G, M, I)

1. $C := \{(G, G')\}$
2. $E := \varnothing$;
3. *currentLevel* $:= \{(G, G')\}$
4. **while***currentLevel* $\neq \varnothing$
5.   *nextLevel* $:= \varnothing$
6.   **for** each $(X, Y) \in$ *currentLevel*
7.   *lowerNeighbours* := *FindLowerNeighbours*$((X, Y))$
8.   **for** each $(X_1, Y_1) \in$ *lowerNeighbours*
9.   **if** $(X_1, Y_1) \notin C$ **then**
10. $C := C \cup \{(X_1, Y_1)\}$
11.   *nextLevel* := *nextLevel* $\cup \{(X_1, Y_1)\}$
12.   Add edge $(X_1, Y_1) \to (X, Y)$ to E
13.   *currentLevel* := *nextLevel*

function *FindLowerNeighbours*$((X, Y))$
/* Returns the lower neighbours of a concept */
1. *candidates* $:= \varnothing$
2. **for** each $m \in M \backslash Y$
3. $X_1 := (Y \cup \{m\})'$
4. $Y_1 := X_1'$
5.   **if** $(X_1, Y_1) \notin$ *candidates* **then**
6.   *candidates* := *candidates* $\cup \{(X_1, Y_1)\}$
7. **return** maximally general *candidates*

Figure 2.7 Next Neighbours algorithm.

algorithm is still $O(|C||G||M|^2)$. Even for this algorithm it is straightforward to build a dual version, the time complexity of the latter being $O(|C||G|^2|M|)$.

We now show the Next Neighbours algorithm at work. The computation of the first three levels of the concept lattice for the context in Table 1.1 is depicted in Figure 2.8. The operation of the inner function is traced in Figure 2.9 for the case where the parent concept is the top element of the lattice. Obviously, the order in which the concepts are generated at each level by the algorithm may not coincide with the ordering used to draw the lattice in Figure 1.5; even the ordering of the letters in each concept intent is determined by the algorithm and may not follow the strict alphabetical ordering used in Figure 1.5.

Before concluding the illustration of the Next Neighbours algorithm, it is useful to point out that the removal of non-maximal candidates can be done more efficiently than suggested above. Although this does not affect theoretical complexity results, as seen above, it may be important in practice. One method is based on the observation that a candidate Y_1' is maximal if and only if every attribute $h \in Y_1'' \backslash Y$ gives rise to the same $(Y_1 \cup \{h\})'$. Note

C = {(1 2 3 4 5 6 7, e)}
currentLevel = {(1 2 3 4 5 6 7, e)}
nextLevel := ∅
lowerNeighbours((1 2 3 4 5 6 7, e)) = {(2 4 5, e c), (1 2 5, e f), (4 5 6 7, e g), (1 3, e h)}
C = {(1 2 3 4 5 6 7, e), (2 4 5, e c)}
nextLevel = {(2 4 5, e c)}
Add edge (2 4 5, e c) → (1 2 3 4 5 6 7, e)
C = {(1 2 3 4 5 6 7, e), (2 4 5, e c), (1 2 5, e f)}
nextLevel = {(2 4 5, e c), (1 2 5, e f)}
Add edge (1 2 5, e f) → (1 2 3 4 5 6 7, e)
C = {(1 2 3 4 5 6 7, e), (2 4 5, e c), (1 2 5, e f), (4 5 6 7, e g)}
nextLevel = {(2 4 5, e c), (1 2 5, e f), (4 5 6 7, e g)}
Add edge (4 5 6 7, e g) → (1 2 3 4 5 6 7, e)
C = {(1 2 3 4 5 6 7, e), (2 4 5, e c), (1 2 5, e f), (4 5 6 7, e g), (1 3, e h)}
nextLevel = {(2 4 5, e c), (1 2 5, e f), (4 5 6 7, e g), (1 3, e h)}
Add edge (1 3, e h) → (1 2 3 4 5 6 7, e)
currentLevel = {(2 4 5, e c), (1 2 5, e f), (4 5 6 7, e g), (1 3, e h)}
nextLevel := ∅
lowerNeighbours((2 4 5, e c)) = {(2 5, e c f), (4 5, e c g)}
C = {(1 2 3 4 5 6 7, e), (2 4 5, e c), (1 2 5, e f), (4 5 6 7, e g), (1 3, e h), (2 5, e c f)}
nextLevel = {(2 5, e c f)}
Add edge (2 5, e c f) → (2 4 5, e c)
C = {(1 2 3 4 5 6 7, e), (2 4 5, e c), (1 2 5, e f), (4 5 6 7, e g), (1 3, e h), (2 5, e c f), (4 5, e c g)}
nextLevel = {(2 5, e c f), (4 5, e c g)}
Add edge (4 5, e c g) → (2 4 5, e c)
lowerNeighbours((1 2 5, e f)) = {(1 2, e f b) ,(2 5, e f c)}
C = {(1 2 3 4 5 6 7, e), (2 4 5, e c), (1 2 5, e f), (4 5 6 7, e g), (1 3, e h), (2 5, e c f), (4 5, e c g), (1 2, e f b)}
nextLevel = {(2 5, e c f), (4 5, e c g), (1 2, e f b)}
Add edge (1 2, e f b) → (1 2 5, e f)
Add edge (2 5, e f c) → (1 2 5, e f)
lowerNeighbours((4 5 6 7, e g)) = {(4 6 7, e g a) ,(4 5, e g c)}
C = {(1 2 3 4 5 6 7, e), (2 4 5, e c), (1 2 5, e f), (4 5 6 7, e g), (1 3, e h), (2 5, e c f), (4 5, e c g), (1 2, e f b), (4 6 7, e g a)}
nextLevel = {(2 5, e c f), (4 5, e c g), (1 2, e f b), (4 6 7, e g a)}
Add edge (4 6 7, e g a) → (4 5 6 7, e g)
Add edge (4 5, e g c) → (4 5 6 7, e g)
lowerNeighbours((1 3, e h)) = {(1, e h b f), (3, e h d)}
C = {(1 2 3 4 5 6 7, e), (2 4 5, e c), (1 2 5, e f), (4 5 6 7, e g), (1 3, e h), (2 5, e c f), (4 5, e c g), (1 2, e f b), (4 6 7, e g a), (1, e h b f)}
nextLevel = {(2 5, e c f), (4 5, e c g), (1 2, e f b), (4 6 7, e g a), (1, e h b f)}
Add edge (1, e h b f) → (1 3, e h)
C = {(1 2 3 4 5 6 7, e), (2 4 5, e c), (1 2 5, e f), (4 5 6 7, e g), (1 3, e h), (2 5, e c f), (4 5, e c g), (1 2, e f b), (4 6 7, e g a), (1, e h b f), (3, e h d)}
nextLevel = {(2 5, e c f), (4 5, e c g), (1 2, e f b), (4 6 7, e g a), (1, e h b f), (3, e h d)}
Add edge (3, e h d) → (1 3, e h)

Figure 2.8 Step-by-step computation of the first three levels of the concept lattice for the context in Table 1.1 using the Next Neighbours algorithm.

FindLowerNeighbours ((1 2 3 4 5 6 7, e))
a: $(\{e\ a\}', \{e\ a\}'') = (4\ 6\ 7, e\ a\ g)$
b: $(\{e\ b\}', \{e\ b\}'') = (1\ 2, e\ b\ f)$
c: $(\{e\ c\}', \{e\ c\}'') = (2\ 4\ 5, e\ c)$
d: $(\{e\ d\}', \{e\ d\}'') = (3, e\ d\ h)$
f: $(\{e\ f\}', \{e\ f\}'') = (1\ 2\ 5, e\ f)$
g: $(\{e\ g\}', \{e\ g\}'') = (4\ 5\ 6\ 7, e\ g)$
h: $(\{e\ h\}', \{e\ h\}'') = (1\ 3, e\ h)$
i: $(\{e\ i\}', \{e\ i\}'') = (7, e\ a\ g\ i)$
candidates = {(4 6 7, e a g), (1 2, e b f), (2 4 5, e c), (3, e d h), (1 2 5, e f), (4 5 6 7, e g), (1 3, e h), (7, e a g i)}
output = {(2 4 5, e c), (1 2 5, e f), (4 5 6 7, e g), (1 3, e h)}

Figure 2.9 Step-by-step computation of the lower neighbours of the top element of the lattice in Figure 1.5.

that this condition is trivially satisfied if there is only one new attribute, as is often the case. Using this method, it is not necessary to collect all candidates first, but care should be taken to avoid selecting the same maximal candidate multiple times. A related, perhaps more elegant and efficient, technique to test candidates for maximality will be presented below.

Using the concepts to find the edges

In the Next Neighbours algorithm, concepts and edges are created simultaneously due to the construction strategy based on the $\prec$ relation; however, the search for the neighbours of the current concept does not take advantage of the concepts that have already been generated.

An alternative method to build the full concept lattice is to create the whole set of concepts first, and then to set the edges. Although this method may seem less efficient, it has the advantage that the generation of candidate neighbours may be faster, provided that an efficient search structure is used to retrieve the concepts that have been created in the first step.

In Figure 2.10, we show the pseudo-code of the *Concepts Cover* algorithm. The top-level description consists of computing the set of concepts C, e.g., by using the Object Intersections algorithm described earlier, followed by the determination of its covering graph. The *CoveringEdges* function works as follows.

One concept (X, Y) of C at a time is examined, and its lower neighbours (or upper neighbours, in the dual version) are found. Similar to *FindLowerNeighbours* in Figure 2.7, the candidate neighbours are generated by considering, for each attribute m not contained in Y, the concept (X_1, Y_1), where X_1 is the set of objects containing both the attributes in X and the attribute m, and Y_1 is the intent of the concept having X_1 as extent.

Concepts Cover
Input: Context (G, M, I)
Output: The concept lattice $L = (C, E)$ of (G, M, I)

1. Find C with the Object Intersections algorithm
2. *CoveringEdges* $(C, (G, M, I))$

function *CoveringEdges*$(C, (G, M, I))$
/* Sets the covering edges between the concepts in C */
1. **for** each $(X, Y) \in C$
2. Set *count* of any concept in C to 0
3. **for** each $m \in M \backslash Y$
4. $inters := X \cap \{m\}'$
5. Find $(X_1, Y_1) \in C$ such that $X_1 = inters$
6. $count(X_1, Y_1) := count(X_1, Y_1) + 1$
7. **if** $(|Y_1| - |Y|) = count(X_1, Y_1)$ **then**
8. Add edge $(X_1, Y_1) \to (X, Y)$ to E

Figure 2.10 Concepts Cover algorithm.

In this case, however, X_1 is found by taking the intersection of X and the attribute extent of m, which is faster than the corresponding operation on line 3 of function *FindLowerNeighbours* in Figure 2.7. Furthermore, rather than computing Y_1 from X_1 via the prime operator, Y_1 is more efficiently retrieved from the concept trie using X_1 as the key.

The problem of selecting maximal candidates is also dealt with in a different way. The algorithm maintains a counter for each concept in C. At the outset each counter is initialized to zero. The counter of a concept (X_1, Y_1) increases whenever that concept is retrieved as a candidate; as soon as the difference between the cardinalities of Y_1 and Y becomes equal to the counter of (X_1, Y_1), the concept (X_1, Y_1) is a child of (X, Y). This ensures that only maximal candidates are selected and that each maximal candidate is taken exactly once.

The worst-case time complexity of the Concepts Cover algorithm is more favourable than that of Next Neighbours. The complexity of step 1, using the Object Intersections algorithm, is $O(|G||C||M|)$. The complexity of *CoveringEdges* is $O(|C||M|(|G| + |M|))$, as lines 4, 5 and 7, which take $O(|G| + |M|)$ time, are executed $|C||M|$ times. The overall time complexity is thus $O(|C||M|(|G| + |M|))$.

In Figure 2.11, we illustrate the working of the *CoveringEdges* function by tracing how it sets the edges between the concept (1 3, e h) in the concept lattice in Figure 1.5 and its children.

Figure 2.11 shows that the concept (1, b e f h) is correctly linked to the concept (1 3, e h) only once, although the same edge is generated as a candidate in two distinct iterations, i.e., for the attributes b and f. Even the bottom element of the lattice is considered a candidate; this guarantees the correctness

a: $inters = \{1\ 3\} \cap \{4\ 6\ 7\} = \varnothing$
$Y_1 = \text{a b c d e f g h i}$
$count(\varnothing, \text{a b c d e f g h i}) = 1$
b: $inters = \{1\ 3\} \cap \{1\ 2\} = \{1\}$
$Y_1 = \text{b e f h}$
$count(1, \text{b e f h}) = 1$
c: $inters = \{1\ 3\} \cap \{2\ 4\ 5\} = \varnothing$
$Y_1 = \text{a b c d e f g h i}$
$count(\varnothing, \text{a b c d e f g h i}) = 2$
d: $inters = \{1\ 3\} \cap \{3\} = \{3\}$
$Y_1 = \text{d e h}$
$count(3, \text{d e h}) = 1$
$|Y_1 - Y| = count(3, \text{d e h})$
Add edge (3, d e h) → (1 3, e h)
f: $inters = \{1\ 3\} \cap \{1\ 2\ 5\} = \{1\}$
$Y_1 = \text{b e f h}$
$count(1, \text{b e f h}) = 2$
$|Y_1 - Y| = count(1, \text{b e f h})$
Add edge (1, b e f h) → (1 3, e h)
g: $inters = \{1\ 3\} \cap \{4\ 5\ 6\ 7\} = \varnothing$
$Y_1 = \text{a b c d e f g h i}$
$count(\varnothing, \text{a b c d e f g h i}) = 3$
i: $inters = \{1\ 3\} \cap \{7\} = \varnothing$
$Y_1 = \text{a b c d e f g h i}$
$count(\varnothing, \text{a b c d e f g h i}) = 4$

Figure 2.11 Step-by-step computation of the lower edges of the concept (1 3, e h) in Figure 1.5 by the inner **for** in the *CoveringEdges* function.

of the algorithm for the case where the concept is to be linked to the bottom element, but it may result in many unnecessary tests for most examined concepts, as in Figure 2.11. A simple alternative is to skip the execution of lines 5, 6 and 7 if *inters* is empty. In this case, at the end of the execution all the concepts without children must be linked to the bottom element.

2.1.4 Construction of partial concept lattices

So far we have considered building the whole concept lattice of a given context. It turns out that in many applications it is enough to compute a very small portion of the lattice, usually consisting of a focus concept and its immediate neighbours.

Such a focus concept, for instance, might be selected by the user through a point-and-click graphical user interface showing a partial lattice, or it might be computed by mapping a natural language or Boolean query on the lattice structure. As a result, the system is supposed to output the neighbours of the selected concepts, which may act as refinements of the concept itself. In Chapters 3 and 4 we will see several applications of this kind.

Here we address the problem of generating all the nearest neighbours of a given concept. The nearest neighbours are formed by the lower and upper neighbours, which can be computed separately.

We have already described in Figure 2.7 one method to determine the set of lower neighbours of a given concept. The algorithm described here follows the same general strategy but differs in two main details, namely the generation of the candidate extent and the choice of the admissible candidates, for which the more efficient procedures presented in Figure 2.10 will be used.

The determination of the upper neighbours is a dual problem and can be easily solved by adapting the function for finding the lower neighbours. In Figure 2.12 we report the complete algorithm.

NearestNeighbours
Input: Context (G, M, I), concept (X, Y) of context (G, M, I)
Output: The set of lower and upper neighbours of (X, Y) in the concept lattice of (G, M, I)

/* Returns the lower neighbours of (X, Y) */
1. *lowerNeighbours* $:= \varnothing$
2. *lNCandidates* $:= \varnothing$
3. **for** each $m \in M \backslash Y$
4. $\quad X_1 := X \cap \{m\}'$
5. $\quad Y_1 := X_1'$
6. $\quad$ **if** $(X_1, Y_1) \notin$ *lNCandidates* **then**
7. $\quad\quad$ Add (X_1, Y_1) to *lNCandidates*
8. $\quad\quad$ $count(X_1, Y_1) := 1$
$\quad\quad$ **else**
9. $\quad\quad$ $count(X_1, Y_1) := count(X_1, Y_1) + 1$
10. $\quad$ **if** $(|Y_1| - |Y|) = count(X_1, Y_1)$ **then**
11. $\quad\quad$ Add (X_1, Y_1) to *lowerNeighbours*

/* Returns the upper neighbours of (X, Y) */
1. *upperNeighbours* $:= \varnothing$
2. *uNCandidates* $:= \varnothing$
3. **for** each $g \in G \backslash X$
4. $\quad Y_2 := Y \cap \{g\}'$
5. $\quad X_2 := Y_2'$
6. $\quad$ **if** $(X_2, Y_2) \notin$ *uNCandidates* **then**
7. $\quad\quad$ Add (X_2, Y_2) to *uNCandidates*
8. $\quad\quad$ $count(X_2, Y_2) := 1$
$\quad\quad$ **else**
9. $\quad\quad$ $count(X_2, Y_2) := count(X_2, Y_2) + 1$
10. $\quad$ **if** $(|X_2| - |X|) = count(X_2, Y_2)$ **then**
11. $\quad\quad$ Add (X_2, Y_2) to *upperNeighbours*

Figure 2.12 Nearest Neighbours algorithm.

The theoretical time complexity of this algorithm is obtained by summing the time taken to find the lower neighbours, $O(|G||M|^2)$, and the time taken to find the upper neighbours, $O(|G|^2|M|)$. In all we get $O(|G||M|(|G| + |M|))$.

The *Nearest Neighbours* algorithm may be iteratively applied to each of the nearest neighbours to find more distant concepts. As the average number of nearest neighbours is usually limited, this is a viable procedure even to generate a relatively large portion of the lattice centred around a focus concept.

Clearly, from an order-theoretic point of view, a partial concept lattice will form neither a lattice nor a sublattice; it will be an ordered set containing a subset of the concepts in the full lattice on which is defined an order relation inherited from the lattice itself.

2.2 Incremental lattice update

In many applications, data are acquired incrementally. One desirable feature to cope with this situation is that the construction of the concept lattice is incremental; i.e., the processing of the nth object does not require extensive reprocessing of the $n - 1$ objects already seen.

Furthermore, to handle dynamic databases more effectively it may be useful to consider not only a change in the set of objects but also changes in the set of attributes. In multimedia systems, for instance, a variety of indices may be used to retrieve information of interest. If the concept lattice is used as a support structure, the user should be given the possibility of dynamically selecting the best attributes according to her goal and profile, as well as to the result of the past interactions with the system and to the given computational constraints.

In this section we will address the issue of the efficient updating of concept lattices, avoiding the need to recompute the whole structure from scratch. We will consider two main tasks: (i) incremental lattice construction as new objects are added, and (ii) incremental lattice updating in response to general variations in the context table. We will then summarize the whole discussion about lattice construction.

2.2.1 Incremental construction of concept lattices

We start by giving an informal characterization of how a concept lattice changes as a new object is added to the context from which it was built. First we note that existing concepts are never removed from the lattice. The extent of an existing concept may contain the new object (i.e., its intent is more general than or equal to the new object's intent), in which case the concept's extent is augmented by the new object. If the extent of an existing concept does not contain the new object (i.e., its intent is incomparable with or is a superset of the new object's intent), then the concept is not affected at all by the object's introduction.

In addition to modifying the extent of the current concepts in which an object is contained, the new object may cause new concepts to be added to the lattice. This happens whenever the intersection of the new object with any set of objects with which it shares some attribute is not already present in the lattice. Any new concept introduced in the lattice must be consistently linked to the other concepts, including other new concepts created by an object's introduction. This will, in general, cause the elimination of some edges (i.e., the edges between all pairs of concepts C_1 and C_2 such that $C_1 > C_{new} > C_2$).

To illustrate, consider the concept lattice shown in Figure 1.5. Suppose we expand the original context by introducing the eighth object shown in Table 2.1. The updated concept lattice is shown in Figure 2.13; the new concepts are in bold, the new edges represented as dashed lines.

In this example, none of the old concepts has been modified; consequently, none of the existing edges have been removed. Five new concepts have been added, including the 'firefly' concept. A new top element with an empty description has been created, consistent with the fact that a 'firefly' is an invertebrate. The concepts 'has wings' (f) and 'has wings, can fly' (b f) have been created because it is no longer the case that all animals that have wings, or that have wings and can fly, also have a skeleton. The concept 'produces light' has been added because this property is now uniquely shared by two animals, the lantern fish and firefly. Note that each new concept may be linked to both existing concepts and other new concepts.

To summarize, the incremental determination of the concept lattice poses three main computational problems: (i) generating all new nodes in the lattice,

Table 2.1 The context for vertebrates in Table 1.1 expanded with an invertebrate animal

		breathes in water (a)	can fly (b)	has beak (c)	has hands (d)	has skeleton (e)	has wings (f)	lives in water (g)	is viviparous (h)	produces light (i)
1	Bat		×			×	×		×	
2	Eagle		×	×		×	×			
3	Monkey				×	×			×	
4	Parrot fish	×		×		×		×		
5	Penguin			×		×	×	×		
6	Shark	×				×		×		
7	Lantern fish	×				×		×		×
8	Firefly		×				×			×

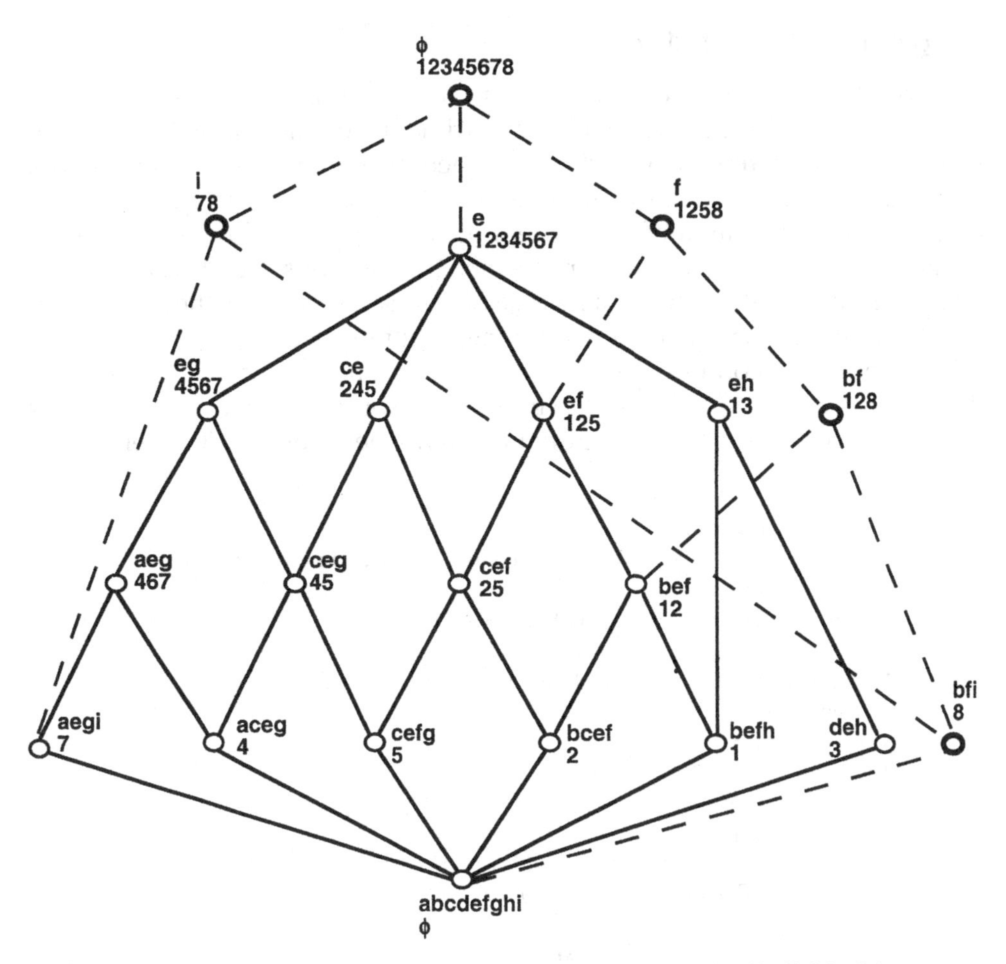

Figure 2.13 Concept lattice for the expanded context in Table 2.1.

(ii) avoiding duplication of nodes that are already present in the lattice, and (iii) updating the edges. As seen earlier for the Object Intersections algorithm, to generate all new nodes it is sufficient to consider the intersections of the new object's intent with each concept's intent in the current lattice; to solve problems (ii) and (iii) different strategies are possible.

In the following we describe two incremental algorithms for building a concept lattice. Each algorithm takes as input a concept lattice and a new object, and returns the updated lattice of the augmented context. For the sake of simplicity, we will first assume that the new object does not contain new attributes with respect to the set of attributes in the original context. Afterwards, we will consider the more general problem of computing an updated lattice when new objects and/or attributes are added to (or possibly deleted from) a given context.

Using concept cardinality

The first algorithm that we present is based on partitioning the set of concepts in the input lattice according to the cardinality of their intents and then processing the partitions, or *buckets*, obtained in ascending cardinality order.

The algorithm, shown in Figure 2.14, iterates over the concepts present in each bucket. If the intent of the concept being examined is contained in or is equal to the new object's intent (line 5), then the concept's extent is augmented by the new object; if the concept has exactly the same intent as the new object, the algorithm stops immediately after the extent update (line 9).

If the previous condition is not met, a new concept is created, provided that it is not already present in the input lattice or in the set of new concepts that have been generated. The extent of such a new concept is given by the

Update by Cardinality

Input: A concept lattice $L = (C, E)$ of context (G, M, I), a new object h with intent $\{h\}'$

Output: The updated concept lattice $L = (C, E)$

1. $oldBucket_j := \{(X, Y) \in C$ such that $|Y| = j\}$
2. $newBucket_j := \varnothing$
3. **for** $j = 0$ **to** $|M|$
4.   **for** each $(X, Y) \in oldBucket_j$
5.   **if** $Y \subseteq h'$
6.   **then**
                   /* modify extent of old concepts */
7.   Replace concept (X, Y) in C with $(X \cup \{h\}, Y)$
8. $newBucket_j := newBucket_j \cup (X, Y)$
9.   **if** $Y = \{h\}'$ **then exit**
10.   **else**
11. $inters = Y \cap \{h\}'$
12.   **if** there is no $(X_1, Y_1) \in newBucket_{|inters|}$ such that $Y_1 = inters$ **then**
                        /* create new concept */
13. $C := C \cup \{(X \cup \{h\}, inters)\}$
14. $newBucket_{|inters|} := newBucket_{|inters|} \cup \{(X \cup \{h\}, inters)\}$
                        /* modify edges */
15.   Add edge $(X, Y) \rightarrow (X \cup \{h\}, inters)$ to E
16.   **for** $k = 0$ **to** $|inters| - 1$
17.   **for** each $(X_1, Y_1) \in newBucket_k$
18.   **if** $Y_1 \subset inters$ **and** there is no (X_2, Y_2) child of (X_1, Y_1) such that $Y_2 \subset inters$ **then**
19.   Add edge $(X \cup \{h\}, inters) \rightarrow (X_1, Y_1)$ to E
20.   **if** (X_1, Y_1) is a parent of (X, Y) **then**
21.   Remove edge $(X, Y) \rightarrow (X_1, Y_1)$ from E
22.   **if** $inters = (h)'$ **then exit**

Figure 2.14 The Update by Cardinality algorithm.

union of the current concept's extent with the new object, the intent by the intersection of the current concept's intent with the new object's intent.

The algorithm maintains a set of new buckets for the concepts that have been modified or created. These new buckets are used to check that the intersection of the current concept with the object is not already present in the lattice (line 12) and are also used to modify the edges.

When a new concept is created, it is linked to the current concept; then its parents are determined by finding those concepts contained in the new buckets such that their intent is a subset of the intersection and their children do not have the same property (line 18). It is also necessary to eliminate an edge between the new parent and the current concept when there is such an edge.

The algorithm halts after processing the bottom element of the lattice, which results in the creation of the new object's concept; it may stop earlier if a concept with an intent containing the new object's intent is encountered (line 22).

One advantage of the *Update by Cardinality* algorithm is that in order to verify that the new concept is not already contained in the lattice it suffices to search the new bucket with the same cardinality as 'intersection', although this could be done even more efficiently by using a search tree. Perhaps more important, the cardinality-based strategy ensures that each new concept can be formed from the new object and the current concept in the simple way described above (see line 13), without the need for computing the prime operator. There is also automatically an edge to one child, and it is not necessary to find the other children of each new concept (if any) because such edges will be created afterwards, as a result of the examination of the concepts with greater cardinality.

To see the algorithm at work, consider the problem of updating the concept lattice shown in Figure 1.5 when the eighth object shown in Figure 2.1 is introduced. In Figure 2.15 we illustrate how the Update by Cardinality algorithm constructs the new concept lattice, shown in Figure 2.13.

For each (non-empty) bucket, we trace only the concepts which affect the lattice update. For instance, $oldBucket_2$ contains three other concepts in addition to (1 2 5, e f), namely (2 4 5, c e), (4 5 6 7, e g), and (1 3, e h). However, as their examination returns an empty *inters* and there is a concept with an empty *inters* in $newBucket_0$, the lattice remains unchanged.

Let us turn to the complexity issue. The outer loops of the algorithm (lines 3 and 4) require $O(|C|)$ time. For each concept in the input lattice, a new concept might be created and its links established. Even assuming that the test to check whether each candidate new concept is already present in the lattice is done using a search tree, in order to set the links properly the algorithm still needs to explore all the buckets that have been already created (lines 16 and 17). In the worst case, the algorithm may have to consider all the concepts in the lattice. For each of these concepts, it must check whether its children,

$oldBucket_1 = \{(1\,2\,3\,4\,5\,6\,7,\ \text{e})\}$
$oldBucket_2 = \{(2\,4\,5, \text{c e}), (125, \text{e f}), (4\,5\,6\,7, \text{e g}), (1\,3, \text{e h}),\}$
$oldBucket_3 = \{(4\,6\,7, \text{a e g}), (1\,2, \text{b e f}), (2\,5, \text{c e f}), (4\,5, \text{c e g}), (3, \text{d e h})\}$
$oldBucket_4 = \{(4, \text{a c e g}), (7, \text{a e g i}), (2, \text{b c e f}), (1, \text{b e f h}), (5, \text{c e f g})\}$
$oldBucket_9 = \{(\varnothing, \text{a b c d e f g h i})\}$

$j = 1$
  $(X, Y) = (1234567, \text{e})$
    $inters = \varnothing$
    new concept $(1\,2\,3\,4\,5\,6\,7\,8, \varnothing)$
    $newBucket_0 = \{(1\,2\,3\,4\,5\,6\,7\,8, \varnothing)\}$
    new edge $(1\,2\,3\,4\,5\,6\,7, \text{e}) \rightarrow (1\,2\,3\,4\,5\,6\,7\,8, \varnothing)$
$j = 2$
  $(X, Y) = (125, \text{e f})$
    $inters = (\text{f})$
    new concept $(1\,2\,5\,8, \text{f})$
    $newBucket_1 = \{(1\,2\,5\,8, \text{f})\}$
    new edge $(1\,2\,5, \text{e f}) \rightarrow (1\,2\,5\,8, \text{f})$
    new edge $(1\,2\,5\,8, \text{f}) \rightarrow (1\,2\,3\,4\,5\,6\,7\,8, \varnothing)$ /* line 19 */
$j = 3$
  $(X, Y) = (12, \text{b e f})$
    $inters = (\text{b f})$
    new concept $(1\,2\,8, \text{b f})$
    $newBucket_2 = \{(1\,2\,8, \text{b f})\}$
    new edge $(1\,2, \text{b e f}) \rightarrow (1\,2\,8, \text{b f})$
    new edge $(1\,2\,8, \text{b f}) \rightarrow (1\,2\,5\,8, \text{f})$
$j = 4$
  $(X, Y) = (7, \text{a e g i})$
    $inters = (\text{i})$
    new concept $(7\,8, \text{i})$
    $newBucket_1 = \{(1\,2\,5\,8, \text{f}), (78, \text{i})\}$
    new edge $(7, \text{a e g i}) \rightarrow (7\,8, \text{i})$
    new edge $(7\,8, \text{i}) \rightarrow (1\,2\,3\,4\,5\,6\,7\,8, \varnothing)$
$j = 9$
  $(X, Y) = (\varnothing, \text{a b c d e f g h i})$
    $inters = (\text{b f i})$
    new concept $(8, \text{b f i})$
    $newBucket_3 = \{(8, \text{b f i})\}$
    new edge $(\varnothing, \text{a b c d e f g h i}) \rightarrow (8, \text{b f i})$
    new edge $(8, \text{b f i}) \rightarrow (7\,8, \text{i})$
    new edge $(8, \text{b f i}) \rightarrow (1\,2\,8, \text{b f})$

Figure 2.15 Computation of the concept lattice for the expanded animal context from the concept lattice of the vertebrates context using the *Update by Cardinality* algorithm.

which are at most $|M|$, are not contained in *inters*, which takes $O(|M|)$. Thus, the time complexity of the algorithm is $O(|C|^2|M|^2)$.

We must emphasize that although the worst-case time complexity grows quadratically with respect to the lattice size, in practice the situation is more contained, because the linking process takes place for only a fraction of concepts and involves checking only a subset of the lattice.

Using local lattice structure

The second incremental algorithm that we describe is based on exploiting the local structure of the lattice to decide when to create a new concept and link it.

Like the Update by Cardinality algorithm, it iterates over the concepts in the input lattice and places edges appropriately. However, the concepts in the current lattice are not examined independently of the way in which they are linked to each other.

The parents of each concept are compared to the intersection of intents of the concept and the new object. If there is a parent with an intent that equals the intersection, then the new concept will not be created because it is already present in the lattice. If there is a parent with an intent that is a proper superset of the intersection, then the new concept will not be created because it will be generated when we intersect the new object with that parent of the node or with one of its ancestors. Otherwise (i.e., if all parents have intents that are proper subsets of the intersection or if all parents are incomparable to the intersection), a new concept will be added to the lattice.

The main function of the *Update by Local Structure* algorithm is shown in Figure 2.16. As in this algorithm the set of concepts is not used separately from their edges, for the sake of simplicity we refer to the lattice as a single entity L, instead of $L = (C, E)$. The lattice is understood as a set of concept records, with fields for extent and intent and pointers to concept neighbours. In Figure 2.16, the main iteration over the set of (old) concepts is preceded by an initialization step that places the new object's concept into the lattice (lines 1–9).

For the problem of setting the appropriate edges, the *LinkConcept* function used by the algorithm operates on the lattice that exists at the time a new concept is created (i.e., this lattice contains all concepts that existed prior to the introduction of a new object, and new concepts that have been created and linked since the new object was first introduced). For each new concept, it determines two boundary sets: the lower boundary contains the most general concepts that are more specific than the new concept, and the upper boundary contains the most specific concepts more general than the new concept. Then the algorithm removes the edges between the two boundaries (if any), and links the new concept to each element in the lower and upper boundary.

Compared to the *FindLowerNeighbour* function introduced in Section 2.1.3, *LinkConcept* has an additional parameter specifying a set of concepts (called descendants) of which the neighbours sought must be seen as ancestors. At the limit, the descendants reduce to the bottom element of the lattice, as when

Update by Local Structure
Input: A concept lattice L of context (G, M, I), a new object h with intent $\{h\}'$
Output: The updated concept lattice L

```
    /* adjust new object's concept */
1.  if there is (X1, Y1) ∈ L such that Y1 = {h}' then
2.      Add h to the extent of (X1, Y1) and of its ancestors in L
3.      exit
4.  oldL := L
5.  if (G, G') ∈ L is such that G' ⊄ {h}' then
6.      Add concept (G ∪ {h}, ∅) to L
7.      Add edge (G, G') → (G ∪ {h}, ∅) to L
8.  Add concept (h, {h}') to L
9.  LinkConcept((h, {h}'), {(M', M)}, L)
    /* iterate over old concepts */
10. for each (X, Y) ∈ oldL
11.     if Y ⊆ {h}' then
12.         Replace concept (X, Y) in L with (X ∪ {h}, Y)
13.     inters = Y ∩ {h}'
14.     if not (
            /* defer creation of new concept or avoid duplicates */
15.         (inters = ∅) or
16.         (inters = {h}') or
17.         (inters = Y) or
18.         (there is an (X2, Y2) parent of (X, Y) in L such that Y2 ⊃ inters))
19.         then
20.             Add concept (X ∪ {h}, inters) to L
21.             LinkConcept ((X ∪ {h}, inters), {(h, {h}'), (X, Y)}, L)
```

Figure 2.16 Main function of the Update by Local Structure algorithm.

the *LinkConcept* function is invoked to place the edges of the new object's concept (line 9 in Figure 2.16).

However, it is sometimes possible to specify some effective descendants, thus restricting the search to a subset of the entire current concept lattice. The process of linking a new concept to its neighbours is a case in point. It can be easily seen that, as the new concept is the intersection of the new object with an old concept, the concepts to which it may be linked must be more general than the new object or more general than the old concept.

In practice, it is convenient to determine either of the two boundary sets, and then derive the other from it; the *LinkConcept* procedure determines the lower boundary first. In the determination of the lower boundary, the algorithm starts from the descendants and keeps visiting ancestors until some concept is found such that its intent contains the new concept's intent. For the upper boundary, the algorithm finds the most specific ancestors of the lower boundary such that their intent is contained in the new concept's intent.

The *LinkConcept* function is detailed in Figure 2.17. In the practical implementation, each ancestor is visited only once through appropriate marking. Similar to the Update by Cardinality algorithm, there is automatically an edge between the current concept and the new concept, without the need for visiting the ancestors of the former. Even the *FindNextLevel* function can be made more efficient by halting the expansion of any branch as soon as a concept is reached that has no attributes in common with the new object's concept.

In summary, the main advantages of the Update by Local Structure algorithm are that (a) each new concept is created only once, and (b) the corresponding links are determined by exploring only the concepts above the

function *LinkConcept* ((X, Y), *descendants, lattice*)
/* Link concept (X, Y) to its neighbours subject to descendants */
1. *lowBoundary* := *FindLowBoundary*((X, Y), *descendants, lattice*)
2. *upBoundary* := *FindUpBoundary*((X, Y), *lowBoundary, lattice*)
3. Remove edges between *lowBoundary* and *upBoundary*, if any
4. Add edges between any concept in *lowBoundary* and (X, Y)
5. Add edges between (X, Y) and any concept in *upBoundary*

function *FindLowBoundary*((X, Y), *descendants, lattice*)
/* Find lower neighbours of concept (X, Y) starting from *descendants* */
1. *lowBoundary* := $\varnothing$
2. **for** each *descendant* $\in$ *descendants*
3.   *lowCand* := $\varnothing$
4.   **for** each (X_1, Y_1) parent of *descendant* in *lattice*
5.   **if** $Y_1 \supset Y$ **then**
6.   *lowCand* := *lowCand* $\cup \{(X_1, Y_1)\}$
7.   **if** *lowCand* = $\varnothing$
8.   **then** *lowBoundary* := *lowBoundary* $\cup$ *descendant*
9.   **else** *lowBoundary* := *lowBoundary* $\cup$ *FindLowBoundary*((X, Y), *lowCand, lattice*)
10. **return** *lowBoundary*

function *FindUpBoundary*((X, Y), *lowBoundary, lattice*)
/* Find upper neighbours of concept (X, Y) starting from *lowBoundary* */
1. *firstLevel* := {parents of *lowBoundary* concepts in *lattice*}
2. *upCand* := *FindNextLevel* ((X, Y), *firstLevel, lattice*)
3. **return** maximally specific concepts in *upCand*

function *FindNextLevel* ((X, Y), *currLevel, lattice*)
1. *upCand* := $\varnothing$
2. **for** each $(X_2, Y_2) \in$ *currLevel*
3.   **if** $Y_2 \subset Y$
4.   **then** *upCand* := *upCand* $\cup\{(X_2, Y_2)\}$
5.   **else** *upCand* := *upCand* $\cup$ *FindNextLevel* ((X, Y), *parents of* (X_2, Y_2), *lattice*)
6. **return** *upCand*

Figure 2.17 Functions used by the Update by Local Structure algorithm to set the edges.

new object's concept (as opposed, for instance, to searching the set containing all the concepts with lower cardinality). It should also be noted that if the concepts were generated in a top-down fashion, as in the Update by Cardinality algorithm, it would be possible to simplify the link process further by generating only one boundary per concept, thus avoiding edge removal altogether.

The algorithm is illustrated using the expanded vertebrates context again. In Figure 2.18 we show how the concept lattice shown in Figure 1.5 is updated when the eighth object shown in Figure 2.1 is added. The result is the lattice shown in Figure 2.13.

The example shows that the algorithm deals efficiently with the concept duplication issue by creating a new concept only when necessary. For instance, the examination of the four concepts (5, c e f g), (2 5, c e f), (2, b c e f), and (1, b e f h) does not lead to the introduction of new concepts, although such concepts have a non-empty intersection with the new object.

As new concepts, when created, are linked to the concepts in the current lattice, the new edges placed by the algorithm depend on which concepts have already been seen. For instance, the edge between (1 2 8, b f) and (1 2 3 4 5 6 7 8, ∅) has been placed because at the time when the new concept (1 2 8, b f) was created the new concept (1 2 5 8, f) had not already been generated, due to the specific order in which the old concepts are examined in the example. Examining the concepts in a different order would produce different temporary edges, but the same final lattice.

The time complexity of the algorithm is given by the product of the time required to iterate over the old concepts and the time necessary to create a new concept and place its edges. The most demanding operation of the latter is the determination of the two boundary sets, which requires comparing the intent of a pair of concepts (line 5 in function *FindLowBoundary* or line 3 in function *FindNextLevel*) for at most as many times as the number of ancestors of the new object's concept. Choosing the total number of concepts as an upper bound of the number of such ancestors, we get $O(|C|^2|M|)$.

The Update by Local Structure algorithm therefore has a worst-case time complexity quadratic with respect to the number of concepts, identical to the previous incremental algorithm. Even so, the practical performance may be considerably better, for two main reasons. The first is that the function *LinkConcept* is invoked exactly as many times as the number of new (distinct) concepts that need be created, usually a small fraction of the number of old concepts in the lattice. The second is that each application of *LinkConcept* requires examining (part of) the concepts that are above the new object's concept, at most 2^k, where k is the number of attributes describing the new object. This is typically a very small subset of the number of concepts.

In fact, several experimental results have been reported that suggest that the time complexity of this algorithm may grow almost linearly in the number

Adjusting new object's concept:

new concept (1 2 3 4 5 6 7 8, ∅) /* line 6 */
new edge (1 2 3 4 5 6 7, e) → (1 2 3 4 5 6 7 8, ∅) /* line 7 */
new concept (8, b f i) /* line 8 */
lowBoundary = {(∅, a b c d e f g h i)}
upBoundary = {(1 2 3 4 5 6 7 8, ∅)}
new edge (∅, a b c d e f g h i) → (8, b f i)
new edge (8, b f i) → (1 2 3 4 5 6 7 8, ∅) /* temporary edge */

Iteration over old concepts:

(4, a c e g), (4 6 7, a e g), (4 5 6 7, e g), (4 5, c e g), (2 4 5, c e), (3, d e h), (1 3, e h), (1 2 3 4 5 6 7 8, ∅) do not alter the lattice because of test on line 15

(5, c e f g), (2 5, c e f), (2, b c e f), (1, b e f h) do not alter the lattice because of test on line 18

(7, a e g i)
  new concept (7 8, i)
  lowBoundary = {(7, a e g i)}
  upBoundary = {(1 2 3 4 5 6 7 8, ∅)}
  new edge (7, a e g i) → (7 8, i)
  new edge (7 8, i) → (1 2 3 4 5 6 7 8, ∅)

(1 2, b e f)
  new concept (1 2 8, b f)
  lowBoundary = {(1 2, b e f), (8, b f i)}
  upBoundary = {(1 2 3 4 5 6 7 8, ∅)}
  removed edge (8, b f i) → (1 2 3 4 5 6 7 8, ∅) /* line 3 in function *LinkConcept* */
  new edge (1 2, b e f) → (1 2 8, b f)
  new edge (8, b f i) → (1 2 8, b f)
  new edge (1 2 8, b f) → (1 2 3 4 5 6 7 8, ∅) /* temporary edge */

(1 2 5, e f)
  new concept (1 2 5 8, f)
  lowBoundary = {(1 2 5, e f), (1 2 8, b f)}
  upBoundary = {(1 2 3 4 5 6 7 8, ∅)}
  removed edge (1 2 8, b f) → (1 2 3 4 5 6 7 8, ∅)
  new edge (1 2 5, e f) → (1 2 5 8, f)
  new edge (1 2 8, b f) → (1 2 5 8, f)
  new edge (1 2 5 8, f) → (1 2 3 4 5 6 7 8, ∅)

Figure 2.18 Computation of the concept lattice for the expanded animals context from the concept lattice of the vertebrates context using the Update by Local Structure algorithm.

of concepts in the lattice rather than quadratically. We will return to this in Section 2.2.3.

2.2.2 Updating the context

In the previous section we addressed the problem of updating the lattice as new objects are added to the context table. Here we tackle the issue of context change in a more general way. The question to be answered is: if we pass from a context producing a lattice L to a new context producing a lattice L', can the lattice L' be computed from the lattice L without recomputing it from scratch?

We break down the problem of context change into two subproblems, namely context enlargement and context reduction, which can be combined to yield the more general case. In each subproblem we study the modification of attributes or objects separately. Therefore we have, in all, four situations:

(1) adding a new object,
(2) adding a new attribute,
(3) deleting an object,
(4) deleting an attribute.

We have already examined the first case, so now we concentrate on the other three situations.

Thanks to the dual nature of concept lattices, the problem of updating the lattice with a new attribute (case 2) can be solved by adapting the algorithms for adding a new object that have been already seen (case 1). In particular, the algorithm for adding a new attribute can be derived from the Update by Local Structure algorithm described in Figures 2.16 and 2.17.

In Figure 2.19 we show the main function of the Add Attribute algorithm derived by duality from the analogous function in Figure 2.16. The function *LinkConcept* invoked in Figure 2.19 can easily be derived from the analogous function in Figure 2.17 by replacing descendants with ancestors, lower with upper, upper with lower, and parents with children.

The worst-case time complexity of the Add Attribute algorithm is the same as the Update by Local Structure algorithm, i.e., quadratic in the lattice size, except that the comparison between concepts is performed by matching their extents, not their intents. Thus, we get $O(|C|^2|G|)$.

Now we turn to context reduction. We start by studying case (3), namely deleting an object without changing the set of attributes. First of all note that the removal of an object only affects the concepts containing that object. By definition, the extent of a concept contains at least the objects of any of its children. Therefore the deletion of an object may have two effects on the concept. Either its extent will continue to differ from that of its children by some attribute, or it will come to have the same description as one of its

Add Attribute
Input: A concept lattice L of context (G, M, I), a new attribute k with extent $\{k\}'$
Output: The updated concept lattice L

```
    /* adjust new attribute's concept */
1.  if there is (X1, Y1) ∈ L such that X1 = {k}' then
2.      Add k to the intent of (X1, Y1) and of its descendants in L
3.      exit
4.  oldL := L
5.  if (M', M) ∈ L is such that M' ⊄ {k}' then
6.      Add concept (∅, M ∪ {k}) to L
7.      Add edge (∅, M ∪ {k}) → (M', M) to L
8.  Add concept ({k}', k) to L
9.  LinkConcept (({k}', k), {(M', M)}, L)
    /* iterate over old concepts */
10. for each (X, Y) ∈ oldL
11.     if X ⊆ {k}' then
12.         Replace concept (X, Y) in L with (X, Y ∪ {h})
13.     inters = X ∩ {k}'
14.     if not (
            /* defer creation of new concept or avoid duplicates */
15.         (inters = ∅) or
16.         (inters = {k}') or
17.         (inters = X) or
18.         (there is a (X2, Y2) child of (X, Y) in L such that X2 ⊃ inters))
19.         then
20.             Add concept (inters, Y ∪ {k}) to L
21.             LinkConcept ((inters, Y ∪ {k}), {({k}', k), (X, Y)}, L)
```

Figure 2.19 Main function of the Add Attribute algorithm.

children. In the latter case the concept has to be removed and the edges have to be changed consistently.

The algorithm for deleting a concept from the lattice works only with the concept extents. It removes the object from each concept of interest; when there is a child of the concept whose extent differs from that of the concept only by the object removed, it deletes the concept. Deleting a concept implies modifying the edges consistently; this is done by linking the children of the removed concept to its parents, unless there is some intermediate concept. A detailed description is provided in Figure 2.20.

The working of the algorithm is illustrated through the inverse operation of expanding the vertebrates context with a new object. This time we want to update the lattice shown in Figure 2.13, relative to the context in Figure 2.1, assuming that the eighth object has been removed from the context. By doing so, we get back to the lattice of the vertebrates context shown in Figure 1.5.

The complete list of operations performed by the *Remove Object* algorithm to update the lattice is shown in Figure 2.21. The concepts containing object 8

Remove Object
Input: A concept lattice L of context (G, M, I), an object h to be removed
Output: The updated concept lattice L

1. $oldL := L$
2. **for** each $(X, Y) \in oldL$
3. **if** $h \subseteq X$ **then**
4. Replace concept (X, Y) in L with $(X \setminus \{h\}, Y)$
5. **if** there is an (X_1, Y_1) child of (X, Y) in L such that $X \setminus \{h\} = X_1$
6. **then** *ReplaceConcept&Edges* $((X, Y), L)$

function *ReplaceConcept&Edges*$((X, Y), L)$
/* Place new edges between parents and children of (X, Y) */
1. **for** each (X_2, Y_2) parent of (X, Y)
2. **for** each (X_1, Y_1) child of (X, Y)
3. **if** there is no (X_3, Y_3) parent of (X_1, Y_1) such that $X_3 \subset X_2$
4. **then** Add edge $(X_1, Y_1) \rightarrow (X_2, Y_2)$ to L
/* Remove concept (X, Y) and its edges */
5. Remove (X, Y) from L
6. Remove all edges between (X, Y) and its neighbours in L

Figure 2.20 The algorithm for updating a lattice after removing an object from the context table.

(7 8, i)
removed concept: (7 8, i)
removed edges: (7, a e g i) → (7 8, i), (8, b f i) → (7 8, i), (7 8, i) → (1 2 3 4 5 6 7 8, ∅)

(1 2 3 4 5 6 7 8, ∅)
removed concept: (1 2 3 4 5 6 7 8, ∅)
removed edges: (1 2 3 4 5 6 7, e) → (1 2 3 4 5 6 7 8, ∅), (1 2 5 8, f) → (1 2 3 4 5 6 7 8, ∅)

(1 2 8, b f)
new edge (8, b f i) → (1 2 5 8, f) /* temporary edge */
removed concept: (1 2 8, b f)
removed edges: (8, b f i) → (1 2 8, b f), (1 2, b e f) → (1 2 8, b f), (1 2 8, b f) → (1 2 5 8, f)

(1 2 5 8, f)
removed concept: (1 2 5 8, f)
removed edges: (8, b f i) → (1 2 5 8, f), (1 2 5, e f) → (1 2 5 8, f)

(8, b f i)
removed concept: (8, b f i)
removed edge: (∅, a b c d e f g h i) → (8, b f i)

Figure 2.21 Computation of the concept lattice for the unexpanded vertebrates context from the concept lattice of the expanded context using the Remove Object algorithm.

are examined in random order; a different order would produce different temporary edges but the same final result, similar to the Update by Local Structure algorithm.

In this particular example, all the concepts containing the object to be removed happen to be deleted from the lattice because they satisfy the condition on line 5 of the main function of the algorithm. In other situations, some such concepts might simply be updated by removing the object from their extent.

It is also worth noting that examination of the concept (1 2 8, b f) causes a new temporary edge to be created between the concept's child (8, b f i) and the concept's parent (1 2 5 8, f) because the condition on line 3 of function *ReplaceConcept&Edges* is satisfied. Such a temporary edge will be deleted in the next iteration.

The complexity of this algorithm, expressed as the number of comparisons between concepts, is fundamentally linear in the lattice size. Deleting objects is therefore much simpler than adding new objects. In the former case we have only to remove concepts from the lattice, whereas in the latter case we have to take into account the interactions between the concepts we create. The practical advantage is that, in order to place edges appropriately at each iteration, it suffices to check the concepts surrounding the current concept (i.e., its parents, its children, and the parents of its children) rather than inspecting a whole subtree or a large subset of the lattice, as in the incremental algorithms seen earlier.

The last subproblem of context change is attribute deletion (case 4). Just as the attribute-adding algorithm is the dual version of the object-adding algorithm, so too the attribute-deleting algorithm may be obtained from the object-deleting algorithm by duality. It suffices to replace object by attribute, extent by intent, child by parent in the main function of the Remove Object algorithm in Figure 2.20, leaving the *ReplaceConcept&Edges* function unchanged. As the derivation is straightforward, we do not report the modified algorithm.

2.2.3 Summary of lattice construction

The possession of a fast algorithm for computing or updating the underlying concept lattice may be an essential prerequisite in many applications. Rule discovery, document ranking and program analysis are just a few examples. In these cases, the resulting concept lattice is not necessarily meant for visual inspection but forms part of a larger software package.

Even if the full lattice is of modest size and/or only a part of it need be visualized, the algorithmic complexity may be a major concern when fast response times are required, as is the case, for instance, in interactive mining of Web retrieval results.

This raises the issue of the optimal choice of the lattice-building algorithm. To ease cross-comparison, we derived the asymptotic worst-case time complexity of the algorithms that have been presented. In this respect, the best result, $O(|C||M|(|G| + |M|))$, was found using the Concepts Cover algorithm. The Next Neighbours algorithm follows closely, with a time complexity of $O(|C||G||M|^2)$, or $O(|C||M||G|^2)$ in the dual version employing the function *FindUpperNeighbours* instead of *FindLowerNeighbours*.

However, the practical behaviour of an algorithm may significantly differ from its worst-case time complexity, partly due to the presence of possibly large constants in the theoretical upper bound, partly because the contexts which yield the worst case may not occur often in practice.

For example, although the asymptotic running time of both of the incremental algorithms presented earlier increases quadratically with the number of concepts, there is a body of experimental evidence suggesting that the growth may, in practice, be linear ([105], [40]). In fact, such algorithms may be an excellent choice not only for incremental tasks but also for batch lattice construction. This is especially true of sparse contexts, which may clearly favour traversal of the intermediate associated lattices over exhaustive search through the input table.

The practical behaviour of the algorithms may vary greatly depending on the type of database being considered, e.g., whether the context table is sparsely or densely filled. A more useful characterization of their complexity should favour average-case analysis and link the performance to some easily measurable features of the input context, such as the size of G and M, the size of the relation I, the average number of attributes per object, and the size of I relative to the product $|G||M|$, which is related to the previous feature.

Since theoretical results of this kind might be difficult to achieve, one may attempt to assess empirically which algorithms perform best on which type of database. For this purpose, it would be very useful to have a set of well-engineered test databases on which to run rigorous experimental comparisons, similar to other standardization efforts being made in related fields such as machine learning and knowledge discovery in databases. This may be an important step to take for the whole research community of concept analysis to encourage systems implementations and to measure advances.

2.3 Visualization

As already noted, most of the applications based on concept lattices require some form of exploration of the graph diagram on the part of the user. However, forming useful visualizations of graph structures is notoriously difficult due to the conflicting issues of size, layout and legibility on limited screen area. Furthermore, visual edge crossing may be detrimental to the comprehension of graph structures.

One main concern is thus the aesthetic of the graph layout. This objective is usually pursued by reducing visual edge crossing or by promoting visual symmetry when the graph itself has symmetrical properties; however, the final result can be evaluated only by the user. Speed is another important criterion, but it may conflict with the former because minimizing the number of edge crossings is usually a time-consuming process. Concept lattices are no exception, although we can take advantage of geometric representations of specific substructures of lattices such as cubes and diamonds which may have some value for the overall structure ([260], [231]).

Regardless of aesthetic and efficiency considerations, some graphs are just too large to fit on a screen. This is typically the case for concept lattices: except for toy databases, we cannot look at the complete set of concepts and edges on one static display. Interactive or incremental or scaling techniques are necessary.

The common approach is to let the user examine a subset of concepts and edges based on the specific task being performed and on the past interaction with the system. In the concept lattice scenario, the system should be able to show or hide parts of the lattice via interactive specification/manipulation of concepts, or relationships between concepts, or even subsets of attributes.

In this section we present three different ways of visualizing a concept lattice that have been implemented with some variants in several system prototypes.

2.3.1 Hierarchical folders

Because of the difficulties associated with visualizing its graph structure, one approach is to reduce the concept lattice to a simpler structure that lends itself to more understandable visual layouts. A tree is the most obvious choice.

A concept lattice can be easily represented as a tree by taking advantage of the subsumption relation that holds between the concepts. The top element of the lattice becomes the root of the hierarchy and each sequence of concepts in the lattice is associated with a path in the tree. This representation is also called a *tree widget*.

To illustrate, in Figure 2.22 we show a (partial) hierarchical representation of the concept lattice introduced in Figure 1.5. The bottom element of the lattice has been omitted and each node in the tree is labelled in such a way that the corresponding concept's intent can be identified by the path from the node to the root of the tree. The extent of concepts is not displayed explicitly, but the number of objects (animals) associated with each node is reported. Only the descendants of one child of the top element of the lattice in Figure 1.5, namely the concept (2 4 5, c e), are shown in full; the other three children of the top element (denoted by an encircled plus) have not been expanded.

It should be noted that in Figure 2.22 there are two distinct paths leading to the same (5, c e f g) concept, consistent with the fact that it has multiple

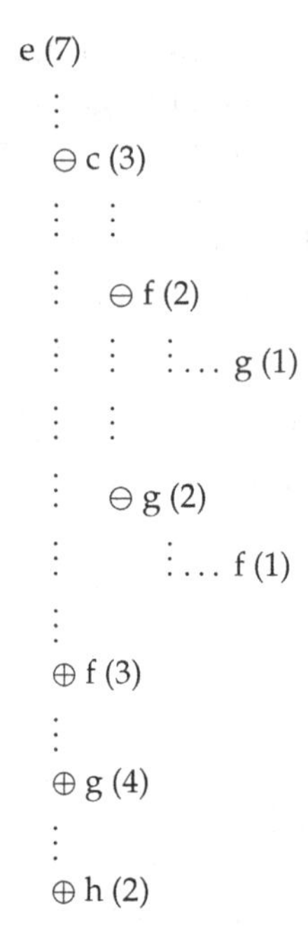

Figure 2.22 Visualization of the vertebrates lattice in Figure 1.5 by hierarchical folders.

parents in Figure 1.5. Indeed, if the hierarchy were displayed in full, the same concept would give rise to as many as four distinct paths.

The tree representation has several advantages. As the metaphor of hierarchical folders is used for storing and retrieving files, bookmarks, menus items, etc., most users are familiar with it and hence no training on the part of the user is required. Furthermore, it takes little space on the screen and it may be drawn efficiently. The main disadvantage, as seen in the example above, is that there may be a considerable amount of duplication of information when the concepts have multiple parents. On the other hand, this is not very likely to happen if only some levels of the hierarchy are visualized, as frequently happens in many applications.

2.3.2 Nested line diagrams

Sometimes it is helpful to split a complex visualization into more individually comprehensible, multiple displays. This is the main rationale behind the technique now described, called *nested line diagrams*. It consists of four main steps:

1. Partition the set of attributes describing a given context in two subsets (for the sake of simplicity we refer to only two subsets, the generalization to more subsets being straightforward).

2. Find concept lattices L_1 and L_2 of the subcontexts identified by the attribute subsets of step 1.
3. Place a copy of L_2 in each node of L_1.
4. Mark with filled circles the elements in each copy of L_2 that belong to the full lattice.

In other words, L_1 is used as an outer frame in which to embed L_2, the copies of which act as a background structure. Mathematically, the full concept lattice is embedded in the direct product of the lattices of subcontexts as a join semilattice. As a result, the concepts and the edges of the full lattice are not directly represented in a nested line diagram. Instead, they can be derived by combining the information associated with the various levels of nesting.

To illustrate, we use the planets context in Table 1.2. Figure 2.23 shows the concept lattice built from two of the three attributes used to describe planets, namely 'distance from sun' and 'moon'. Figure 2.24 shows the combination of the lattice in Figure 2.23 and the lattice built from the third attribute, 'size'. Each line in the nested diagram has to be interpreted as a bundle of parallel lines connecting the respective pairs of elements of the background marked with filled circles.

The intent of a concept in a nested line diagram can be determined by collecting attributes both in the outer frame and within the local diagram. For example, for the leftmost small circle in Figure 2.24 we get *mn*, *dn* from the

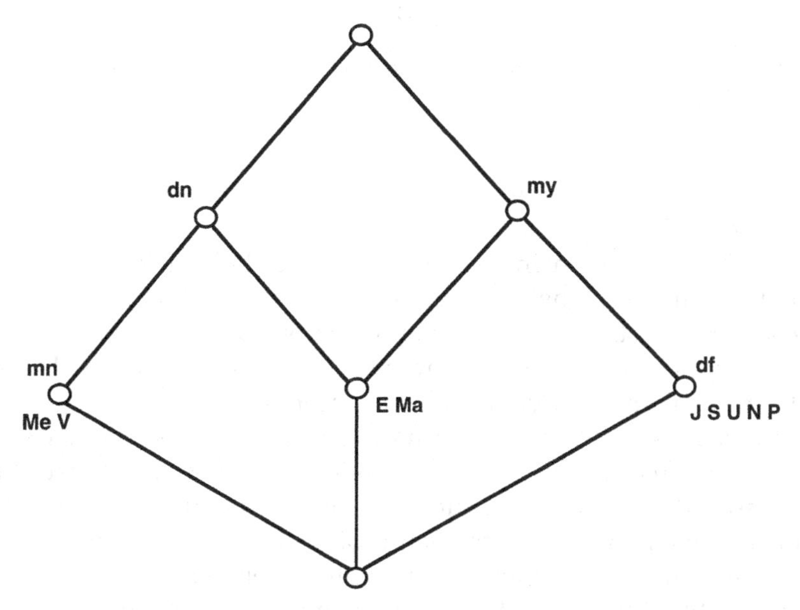

Figure 2.23 Concept lattice of the subcontext 'distance from sun & moon' in the planets context in Table 1.2.

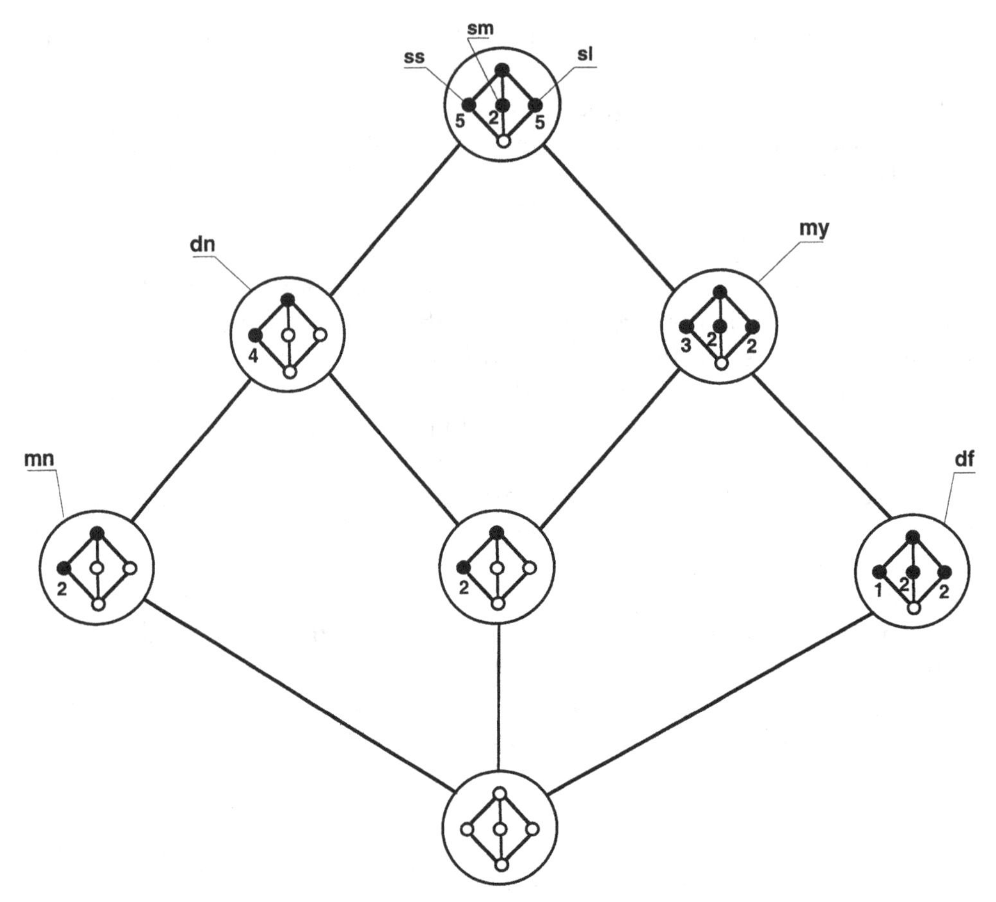

Figure 2.24 A nested line diagram for the two subcontexts 'distance from sun & moon' and 'size'.

outer lattice and *ss* from the inner lattice. The two planets associated with it are Mercury and Venus.

Not all concepts shown in the direct product belong to the lattice. For instance, there are no planets of medium or large size among those that are near the sun. Hence, the corresponding concepts in Figure 2.24, located in the *dn* and *mn* large circles in the outer frame, remain as open circles. These void concepts are sometimes called *unrealized* concepts.

This technique is most suitable for many-valued contexts, as in the example given above, because it may be easier to identify valuable subcontexts. Indeed, in many cases, the lattices of certain subcontexts may be seen as conceptual scales of the given context, in the sense of Section 1.3. In principle, nested line diagrams can also be applied to one-valued contexts by vertically slicing the context table, but in the latter case it may be more difficult to select informative subcontexts.

The metaphor of conceptual scales has been effectively implemented in the *Anaconda* and *Toscana* tools ([247], [248]), which together form a system for the creation and combination of conceptual scales in a nested line diagram. The conceptual scales are created in a first step using Anaconda and are next made available to users for analysing the data through Toscana.

The basic searching mechanism offered by the system is based on *nesting and zooming*: nesting is for inserting the line diagram of a new scale in the nodes of a current outer scale, and zooming allows the user to choose a certain concept of the outer scale and display a differentiation of that concept by means of a nested line diagram. The system is maintained and updated through an ongoing software development project (see http://tockit.sourceforge.net/toscanaj).

One advantage of nested line diagrams is that the size of each local diagram cannot exceed the number of possible combinations of the attribute values present in the corresponding subcontext, regardless of the number of objects in the database. Hence, it is possible to draw the full lattices of each subcontext even for large databases, provided that the subcontexts are sufficiently small.

Clearly, this approach is effective when the number of scales to combine is limited. Furthermore, the complete direct product, as introduced above, has the disadvantage that there may be many unrealized (or 'ghost') concepts in the displayed combination even using few small scales.

One way to reduce the number of unrealized concepts in the visualization is to obtain the semi-direct product of the lattices of subcontexts, which basically amounts to leaving out empty bottom elements in the framed circles. A further refinement, termed *local scaling*, consists of omitting all the inner concepts that are irrelevant to the corresponding facet of the outer frame [224].

Before concluding this section, we wish to emphasize that although the nested line diagram technique originated in the concept lattice community, the idea of combining multiple partial views is also well represented in the information visualization field and in the information science literature. We will return to this in the bibliographic notes at the end of the chapter.

2.3.3 Focus+context views

Unlike nested line diagrams, in which only one or more subsets of the original attributes are used at a time, the techniques discussed in this subsection consider the concept lattice to be built from the whole set of attributes. However, the lattice is not displayed in full, and not all displayed parts are rendered in the same way.

Greater prominence is given to a certain area of the visualization space where information of interest to the user is presumably located. In other terms, it is assumed that there is a current focus of interest in the lattice to be displayed. Such a focus usually corresponds to a certain concept, which may be identified in different ways depending on the application at hand. In

Part II of the book we will see several interaction mechanisms of this kind, including querying, point-and-click selection, and thesaurus climbing.

The main feature of focus+context techniques is the smooth integration of the focus with the surrounding context. In contrast to other approaches based on multiple views such as *local* and *global* views, the disadvantage of which is that the user must map different graphical representations, the transition from focus to context is made more continuous by allowing for variable magnification of graphic displays.

In essence, the information contained in the graph is shown at varying levels of detail depending on the distance from the focus; the amount of information at the focal point is increased, whereas information placed further away is reduced in quantity.

Focus+context viewers use as a general metaphor the effects observed when looking through a fisheye lens or magnifying glass. As an illustrative example, in Figure 2.25 we show what a fisheye view of the concept lattice shown in Figure 1.5 might look like, relative to the focus concept (2, c e f). The bottom element of the lattice has been omitted, whereas the top element of the lattice, described by a non-empty intent, has been included in the visualized display. For each node (concept), its intent is shown along with the cardinality of its extent.

To draw the nodes we used a specific display format for each subset of concepts placed at the same distance from the focus concept, the distance being the length of the shortest paths between the concepts. In particular, we used five types of display involving different combinations of sizes, fonts,

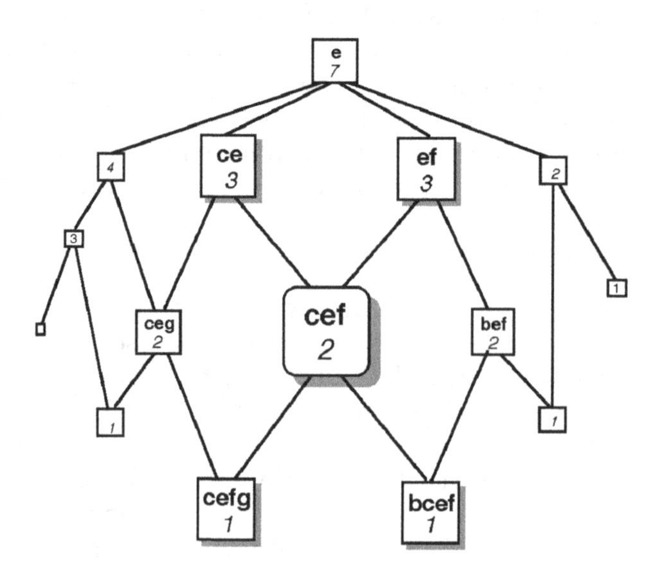

Figure 2.25 Fisheye view of the vertebrates lattice in Figure 1.5, with focus on the concept (2, c e f).

and types of information. The focus node is a large box with large fonts. Its nearest nodes still show full information but with increasingly reduced sizes and fonts. Further nodes are shown as small boxes containing only the number of associated documents. The farthest node shows just the structure's topology.

The metaphor of the fisheye view can also be used at more discrete levels, thus leading to an approximate, yet more efficient, solution to the continuous range effect problem. One simple approach is to rely on a user-supplied function called DOI (Degrees of Interest) associated with objects in the visualized structure. Such a function produces a range of values which can be compared to a threshold value to decide whether the associated objects should be displayed or not. Hiding or condensing clusters of information, termed elision, is another useful technique to create a range of visualizations.

Such discrete techniques may have the same overall effect as truly focus+context viewers, as the removal of unwanted information from the visualization indirectly results in promoting the required information. However, the strategy is different in that they affect the choice of the objects to be visualized rather than the space in which they are visualized. In this respect, these techniques could perhaps be more appropriately described as visual filters, based on a set of conditions controlled by the user. We will see a related approach in Section 2.4.2, where we discuss an interactive pruning mechanism for concept lattices based on a user's constraints.

Following a similar hybrid approach, we now outline a method to implement a fisheye-like view of a concept lattice. The method is based on two steps: the first computes the graph to be drawn, the second draws it.

To compute the graph, one level at a time is generated; the first level is just the focus node, and the nth level is obtained from the $(n - 1)$th level by considering for each node in the $(n - 1)$th level all the adjacent nodes that have not already been generated. Before adding another level to the graph to draw, the algorithm checks that the size of the resulting graph does not exceed a given constant. The size is computed by a weighted sum of all the nodes in the graph, where the weight of each node is inversely proportional to its distance from the focus node.

Also, while computing each level the algorithm checks whether the number of nodes adjacent to each single node to be expanded exceeds a second constant, in which case they are not included in the set of nodes to be drawn (instead, the node to be expanded will be simply labelled with the number of its neighbours when drawing the graph). The algorithm, called *Focus+Context*, is detailed in Figure 2.26.

Once we have decided which nodes to draw using which format – in practice each level is assigned a set of rules, as illustrated above – the actual layout is produced by a standard graphical routine for painting graphs.

Focus + Context

Input: A concept lattice L, a focus concept (X_1, Y_1), a max-weighted-size threshold, a max-nearest-neighbours threshold.

Output: A graph formed of a set C^* of pairs (concept, distance-from-focus) and a set of edges E.

1. $weightedSize := 0$
2. $distance := 0$
3. $currentLevel := \{(X_1, Y_1)\}$
4. $nextLevel := \emptyset$
5. **while** $weightedSize < maxWeightedSize$
6. $\quad distance := distance + 1$
7. $\quad$ **for** each $(X_2, Y_2) \in currentLevel$
8. $\qquad$ **if** the number of nearest neighbours of (X_2, Y_2) is $\leq maxNearestNeighbours$ **then**
9. $\qquad\quad$ **for** each nearest neighbour (X_3, Y_3) of (X_2, Y_2)
10. $\qquad\qquad$ Add edge $(X_2, Y_2) \rightarrow (X_3, Y_3)$ to E
11. $\qquad\qquad$ **if** (X_3, Y_3) has not been marked **then**
11. $\qquad\qquad\quad$ mark (X_3, Y_3)
12. $\qquad\qquad\quad C^* := C^* \cup \{(X_3, Y_3), distance\}$
13. $\qquad\qquad\quad weightedSize := weightedSize + Size((X_3, Y_3))$
14. $\qquad\qquad\quad nextLevel := nextLevel \cup \{(X_3, Y_3)\}$
15. $\qquad currentLevel := nextLevel$

Figure 2.26 An algorithm to find a portion of lattice to visualize according to the focus+context metaphor.

Although such routines may not produce particularly pleasing layouts, they are very simple and efficient to use. If we wanted to produce a more aesthetic and fisheye-oriented layout we would control not only the size of the elements of the graph but also their position, e.g., by using the more sophisticated techniques for producing the continuous range effect that have been mentioned above.

2.4 Adding knowledge to concept lattices

In many applications, a body of background knowledge is available that can be used to better model and analyse the data represented in the context. In Section 2.4.1 we will see how concept data analysis can be extended to cope with this situation.

Another useful source of information not encoded in the context is represented by the knowledge about the goal and/or the domain that the user has or learns during interaction with the lattice. As the number of concepts may grow large in most applications, the user's knowledge can be especially useful for discarding the concepts that are not truly relevant to the task at hand. In Section 2.4.2 we will describe a form of concept pruning driven by the user's knowledge.

2.4.1 Adding background knowledge to object description

One common source of background knowledge is represented by a partial ordering relation $\leq_{M^*}$ between a superset M^* of the attributes describing the objects. We will assume that the intended meaning of the ordering relation is that any attribute implies any of its more general attributes. In other words, we will assume that the following *compatibility condition* holds:

$$\forall g \in G, m_1 \in M, m_2 \in M^* : [(g, m_1) \in I, m_1 \leq_{M^*} m_2] \Rightarrow (g, m_2) \in I.$$

The corresponding condition for the case when an ordering relation $\leq_{M \times V^*}$ holds between a superset V^* of the values taken on by an attribute M is that for each $g \in G, (m, v_1) \in M \times V, (m, v_2) \in M \times V^*$,

$$[(g, m, v_1) \in I, (m, v_1) \leq_{M \times V^*} (m, v_2)] \Rightarrow (g, m, v_2) \in I.$$

Usually, the subsumption relation will represent a tree, but for our purposes this is not necessary: it can be any directed acyclic graph with a unique maximal element. This condition is necessary to guarantee that for each pair of descriptors (attributes or attribute values) there will be an element that subsumes both.

If such a maximal element is not contained in the original relation, we may simply add a new 'dummy' top element. In this way we get a join semilattice, where the top element is understood as the upper bound of those pairs of elements that have no more general terms in common according to the subsumption relation.

The next question is: how is it possible to build a concept lattice from a context and from a subsumption hierarchy over the context's descriptors?

One simple method for taking into account structured descriptors would be to expand the context table that describes the data, adding to the description of each object all the attributes (or attribute values) that are implied by the original attributes. This would amount to creating a conceptual scale of the type seen in Section 1.3.

At this point one could treat the new context as if it contained only unordered descriptors, thus using the methods seen so far. However, the space necessary to store the expanded context would increase considerably and so would the time taken by the algorithms to perform the basic operations involving intent representations.

Here we present a different approach. The idea is to work with an unexpanded context representation and to adapt the construction algorithms to take into account the implicit descriptors. For the sake of simplicity, we will refer to the Update by Local Structure algorithm presented in Section 2.2.1, but similar considerations also apply to most other algorithms because the basic tests and operators affected are the same.

We first discuss the simpler case in which a subsumption hierarchy is defined over the values taken on by a single (many-valued) attribute, and then we cover the use of a subsumption hierarchy defined over the whole set of (one-valued) attributes.

Structuring single many-valued attributes

We will illustrate our approach through a simple example in the playing cards domain.

We consider the two simple attributes: rank = {1, 2, 3, 4, 5, 6, 7, 8, 9, 10, J, Q, K} and suit = {♣, ♠, ♡, ◇}, along with the set of objects with the following object intents: 2♣, 2◇, 5♣, *J*♡, *K*♠. The corresponding concept lattice is illustrated in Figure 2.27.

Rather than using a set of unordered elements, we may assume that the range of each attributes is tree-structured. Two possible trees are shown in Figure 2.28. The leaves of each tree can be thought of as the observable features of the corresponding attribute, whereas the other nodes in the trees are the non-observable features.

Getting back to the problem of extending the Update by Local Structure algorithm to take into account structured attributes, it will immediately be

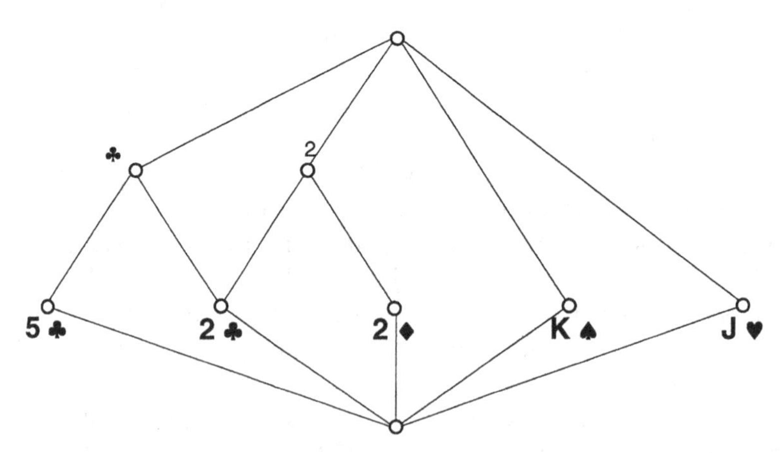

Figure 2.27 The concept lattice for the playing cards context.

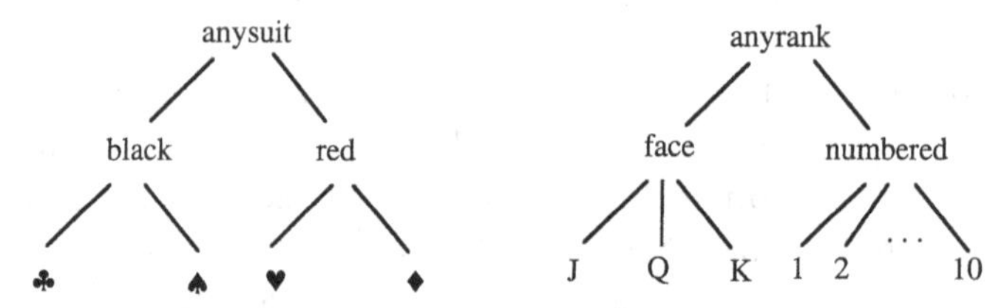

Figure 2.28 Two tree-structured attributes in the playing cards domain.

realized that the two steps affected by the new scenario are the computation of the intersection between concepts and the assessment of the ordering relation $\leq$ between concepts.

The standard set intersection operator can no longer be used to find the intersection between a pair of concepts which may share implicit attribute values. Denote by $\leq_{M_i}$ a partial ordering relation defined over the values taken on by the ith attribute. The intersection of two concepts (X_1, Y_1), (X_2, Y_2) is obtained by finding for each pair $[(m_i, v_j); (m_i, v_k)]$, $(m_i, v_j) \in Y_1, (m_i, v_k) \in Y_2$, the lowest upper bound in the corresponding ordering relation $\leq_{M_i}$.

Similarly, checking whether $(X_1, Y_1) \leq (X_2, Y_2)$ cannot be done using the standard set inclusion relations between Y_1 and Y_2. To account for the implicit attribute values, we need to verify that the relation $\leq_{M_i}$ holds between each attribute--value pair in Y_1 and the corresponding pair in Y_2.

Consider that a tree-structured attribute represents a fair generalization of a nominal attribute, in that the latter can be converted into a tree-structured attribute using a dummy root 'any-value'. Furthermore, the same approach applies to a larger class of structured attributes than trees because the only property we have used is the existence of an upper bound for any pair of attribute values. In fact, as already mentioned, it is sufficient that the ordering relation be a join semilattice.

With the extended version of the algorithm, it is possible to build the expanded lattice directly from the unexpanded context and the subsumption hierarchies. For the playing cards example, we get the concept lattice shown in Figure 2.29.

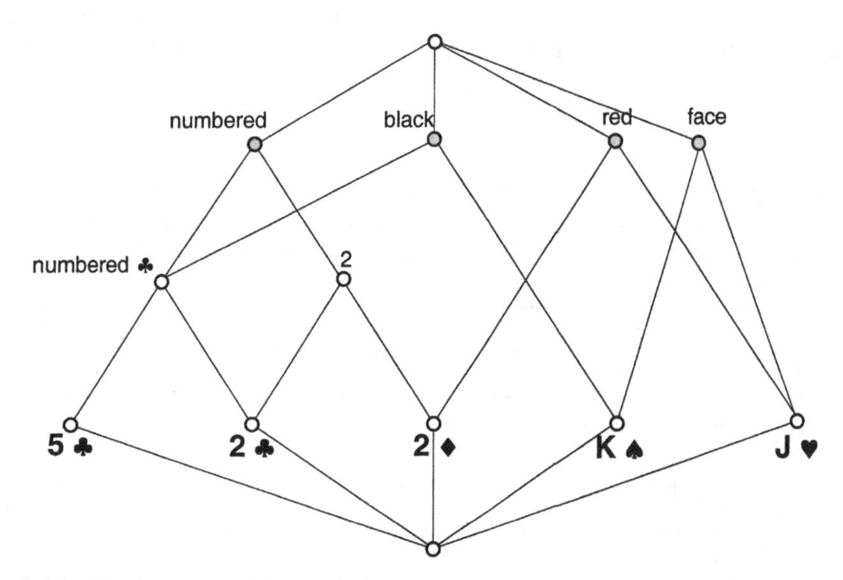

Figure 2.29 The concept lattice for the playing cards context augmented by the background knowledge shown in Figure 2.28.

The concepts in the expanded lattice L^* represent a superset of the concepts in the unexpanded lattice L. While the introduction of new descriptive terms has the effect of revealing similarities between objects that had not been formerly grouped (the grey circles in Figure 2.29), it is interesting to see what happens to the concepts in L. Their extent remains the same, but their intent may change due to the fact that the same set of objects may now be characterized by a different (more specific) intensional description. The concept '♣' in L, for instance, which groups the two objects 2♣ and 5♣, turns into the new concept 'numbered, ♣' in L^*.

Structuring all the one-valued attributes

One-valued contexts with background information are often encountered in information retrieval tasks, where the background knowledge is encoded in a thesaurus of the terms describing the documents.

We will use as an illustration an example presented in [40]. Table 2.2 shows a tiny fragment of the INSPEC computer-engineering collection (http://www.iee.org/Publish/Inspec/) consisting of six objects described by eight attributes, and Figure 2.31 shows the corresponding concept lattice. In Figure 2.30 we show the portion of the 1991 INSPEC thesaurus encoding the broader/narrower relation among the term attributes in Table 2.2.

In this case, the background knowledge is not structured as a tree, because the attribute EXPERT-SYSTEMS has two parents. Also, it covers neither a subset nor a superset of the original attributes.

In fact, the graph contains new attributes, such as INFORMATION-ANALYSIS and COMPUTER-APPLICATIONS , which did not appear in the document descriptions. On the other hand, the attribute ARTIFICIAL-INTELLIGENCE is used to describe the documents but it does not appear in the graph.

Table 2.2 Context of a tiny subset of the INSPEC document collection. The attributes are: ARTIFICIAL-INTELLIGENCE, EXPERT-SYSTEMS, INFORMATION-RETRIEVAL,CATALOGUING, INDEXING, INFORMATION-SCIENCE, INFORMATION-RETRIEVAL-SYSTEMS, KNOWLEDGE-BASED-SYSTEMS

	AI	ES	IR	Cat	Ind	IS	IRS	KBS
d1	x	x	x					
d2	x	x		x				
d3	x	x			x			
d4	x	x				x		
d5	x	x		x			x	
d6	x		x					x

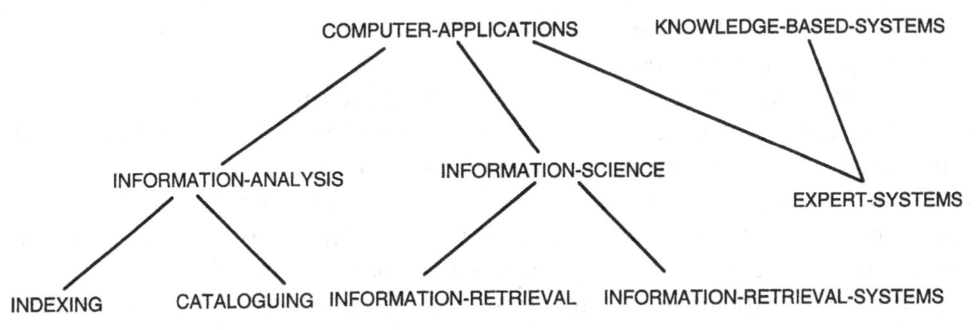

Figure 2.30 Graph-structured background information over the attributes in Table 2.2, taken from (40).

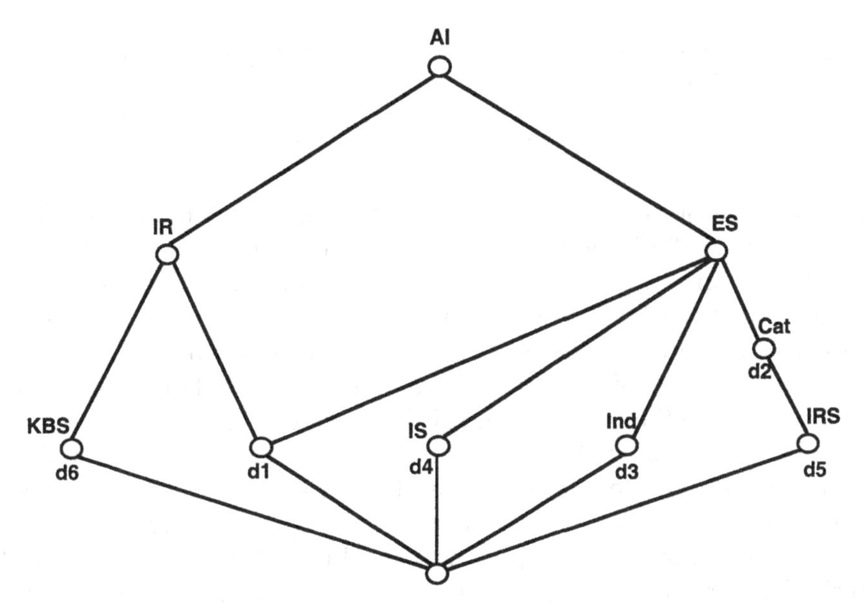

Figure 2.31 The concept lattice for the document context in Table 2.2.

To deal with the latter type of attribute, it is convenient to add a most general attribute ANY to the original relation and include the missing attributes as children of ANY.

The new ordering relation, denoted $\leq_{M^*}$, will cover a superset of M. It can also be seen as a relation defined over a particular many-valued attribute of type content, which may take on the values specified by the original one-valued attributes; such attribute values are then partially ordered according to $\leq_{M^*}$.

Let us now see how the background knowledge affects lattice construction. Handling one-valued contexts is more difficult than many-valued contexts because there may be possible interferences between the upper bounds of different pairs of attributes.

The generalized operator ($\cap^*$) for computing the intersection between two expanded concepts can best be described operationally: the intersection of the intents of two concepts (X_1, Y_1) and (X_2, Y_2) is obtained by finding, for each pair (m_1, m_2), $m_1 \in Y_1$, $m_2 \in Y_2$, the most specific attributes in M^* that are more general than m_1 and m_2, and then retaining only the most specific elements of the set of attributes generated in this way.

To account for the implicit attributes, the definition of the ordering relation $\leq$ between the concepts can be rewritten, for the intent part, in the following way (the definition for the extent part is left unchanged):

$$(X_1, Y_1) \leq (X_2, Y_2) \iff \forall\, m_2 \in Y_2,\ \exists\, m_1 \in Y_1,\ m_1 \leq_{M^*} m_2$$

In practice, it is convenient to assess the relation $\leq$ using the generalized intersection operator $\cap^*$:

$$(X_1, Y_1) \leq (X_2, Y_2) \iff (X_1, Y_1) \cap^* (X_2, Y_2) = (X_2, Y_2)$$

The concept lattice for the augmented context is shown in Figure 2.32.

Even the result of adding background knowledge may be more difficult to interpret for a one-valued context than for a many-valued context because the number of attributes per object is not fixed.

The size of the lattice, when passing from an unstructured to a structured and possibly larger set of attributes, usually increases but it may also decrease. This contrasts with the many-valued single-attributes case.

To explain, consider that while the addition of new attributes is likely to produce new concepts by revealing similarities between sets of objects that were formerly ungrouped (e.g., concept (1 4 5 6) in L^*), attribute structuring alone may reduce the number of concepts. This will happen whenever two formerly distinct concepts come to have the same description after attribute expansion, thus contracting to a single concept.

Attribute structuring may also cause two incomparable concepts to become comparable. In this case the ordering relation changes whereas the number of concepts either decreases or remains the same.

An example of the former situation is represented by concepts (1), (6), and (1 6), which were distinct in L (see Figure 2.31). After the introduction of the background knowledge, object 1 is described by a subset of the attributes describing object 6, and thus object 1's concept coincides with concept (1 6) in the expanded lattice L^*. An example of the latter case is given by concepts (1), (4), (5); they are incomparable in L, whereas the intent of (4) becomes more general than that of both other objects in L^*.

The overall behaviour depends on the specific data, on the new set of attributes, and on the ordering structure over the attributes, although the

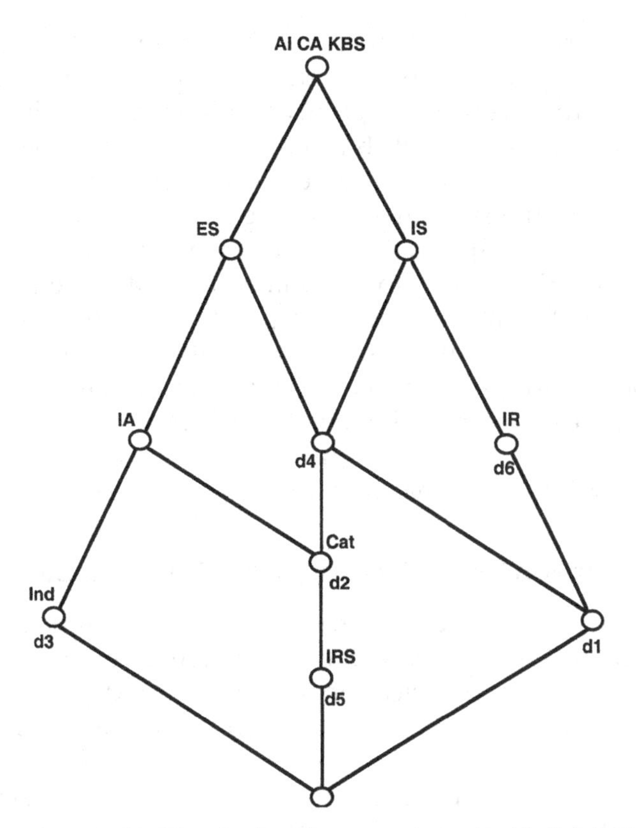

Figure 2.32 The concept lattice for the document context in Table 2.2 augmented by the background knowledge shown in Figure 2.30.

inclusion of background knowledge usually results in a noticeable increase in the lattice size and in a higher degree of structuring, as will be seen in Section 3.1.2; i.e., the expanded lattice may be deeper and narrower than the unexpanded one, as in our example.

Computational complexity of the extended algorithm

We now briefly examine how the time complexity varies when the Update by Local Structure algorithm works with structured descriptors. We will consider the case of single many-valued attributes first.

The theoretical bounds found when the Update by Local Structure algorithm runs with unstructured attributes remain the same. However, those bounds refer to the number of comparisons between concepts and objects and comparisons between concepts, without considering the cost of the comparison itself. In the case of unstructured attributes, each such comparison requires matching the corresponding values of each attribute, therefore the time necessary is only proportional to the number of attributes.

In the case of structured attributes, for each attribute the algorithm must check whether the two values are ordered with respect to the ordering relation defined over the attribute. The complexity of this operation is in general linear in the number of values in the attribute range, or smaller in special cases (e.g., logarithmic for tree-structured attributes). Therefore the time complexity of each comparison in the structured case is greater than in the unstructured case by a factor equal to the size of each attribute's range.

Similar considerations apply to one-valued contexts. The main difference is that, for each attribute in one concept, the algorithm must check its descendants in $\leq_{M^*}$ to see if it can find *some* attribute of the other concept. In fact, by using mark propagation, this problem can be solved by visiting all the elements below each attribute only once. Hence, analogous to many-valued contexts, the time complexity in the structured case is greater than the unstructured case by a factor that is, at most, equal to the number of attributes.

Utility of subsumption hierarchies for lattice-based applications

Before concluding this subsection, we wish to briefly address the utility issue. As will be better seen in Part II, the integration of existing subsumption hierarchies into a concept lattice modelling a context is an essential step for several applications, most notably for information retrieval and software engineering.

For information retrieval purposes, the use of a thesaurus basically makes it possible to create new meaningful queries and guarantees that more general queries are indexed with more general terms, whereas in an unexpanded concept lattice each query is strictly described by the terms present in the documents, and possible semantic relationships between the terms themselves are ignored.

As a result, the user may locate information of interest more quickly by means of focused queries, and the overall retrieval effectiveness of lattice navigation may considerably improve. This will be discussed in Section 3.1.2.

For software engineering applications, the basic technique is the same but the interpretation and utilization of the results is quite different. Rather than improving the representation of the context at hand, the background knowledge is integrated into the lattice with the goal of analysing the appropriateness of the background knowledge itself in a specific usage situation.

The best-known example is the analysis of class hierarchies in object-oriented software [215]. The expanded concept lattice may be used to find out which classes should be removed, or moved into a derived class, or split into multiple classes, to better reflect their use in a given set of applications.

A similar approach can be taken to analyse the behaviour of a classification thesaurus with respect to certain document collections. This application will be specifically addressed in Section 4.1.

In Part II, we will see several examples of the utilization of a subsumption hierarchy as a performance component of a larger lattice-based program.

2.4.2 Pruning concepts with user constraints

One natural and powerful form of concept pruning consists of letting the user incrementally restrict the set of concepts being inspected using her knowledge about the goal and/or the domain.

Here we define a simple framework whereby the user may apply constraints with which the concepts sought have to comply and have the system prune the search space accordingly. The framework guarantees that as more and more constraints are specified the search space will converge to the target concept(s).

The constraints may be expressed as order-theoretic operators defined over a specific subset of attributes. For the sake of illustration we will assume that such attributes coincide with the intent of some concept c_1 in the lattice, but this is not strictly necessary; any given subset of attributes may induce a set of satisfying concepts.

Denoting the unpruned concept lattice by $(C; \leq)$, we define four types of constraints:

1) the set of elements up c_1, i.e., $\uparrow c_1 := \{c \in C | c_1 \leq c\}$,
2) the set of elements down c_1, i.e., $\downarrow c_1 := \{c \in C | c \leq c_1\}$,
3) the complement of up c_1, i.e., $\neg \uparrow c_1 := \{c \in C | c_1 \neg \leq c\}$,
4) the complement of down c_1, i.e., $\neg \downarrow c_1 := \{c \in C | c \neg \leq c_1\}$.

These constraints have an immediate graphic interpretation in terms of the partitions they induce over the search space, as shown in Figure 2.33; they cause the white regions to be pruned away from the space, thus restricting the admissible space to the grey regions. Note that the admissible regions corresponding to constraints of type $\uparrow c_1$ and $\downarrow c_1$ are sublattices of $(C; \leq)$.

The four constraints also allow a nice interpretation from the point of view of the information retrieval properties of the induced partitions; we will return to this aspect in Section 4.1.

Of course, this is an abstract definition of the pruning framework. The next step is to see whether there is a compact and computable representation for the pruned space beyond simple listing of all admissible classes, and whether such a representation supports efficient updating.

To address this issue, one can build on ideas developed in the machine learning field for inducing concepts from examples and counter-examples ([167], [158]); in fact, it turns out that while these ideas have been developed for an entirely different task they can be extended and adapted to solve our problem.

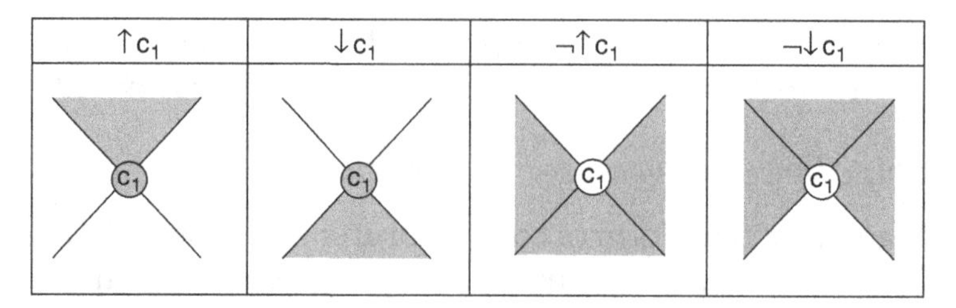

Figure 2.33 Pictorial representation of the user constraints.

To start with, the search space at any given time is represented by two boundary sets, S and G, where S contains the set of most specific consistent concepts (by 'consistent' we mean that all seen constraints are satisfied) and G contains the set of most general consistent concepts. A concept c is admissible if there is some $s \in S$ such that $s \leq c$, and some $g \in G$ such that $c \leq g$. This is called a version space representation.

Constraints of type $\uparrow$ and $\neg \downarrow$ will move S upward, whereas constraints of type $\downarrow$ and $\neg \uparrow$ will move G downward; if/when the two boundary sets become equal they will contain just the target concept(s).

Using a version space representation, the user constraints can be examined in an incremental fashion, without recomputing the pruned space from scratch whenever a new constraint is added. In Figure 2.34 we show how to find the elements of the new S and G sets from the current S and G sets. Note that min and max stand for minimal and maximal elements (see Section 1.1).

It is easy to show that if $(C; \leq)$ is finite the framework is theoretically correct, i.e., it is sound and complete, and the result does not depend on the order of presentation of constraints. Furthermore, if direct access to each concept's

$\uparrow c_1$:

$$G_{k+1} = \{g \in G_k | c_1 \leq g\}$$
$$S_{k+1} = \min\{c \in C | c_1 \leq c, \exists s \in S_k : s \leq c, \exists g \in G_{k+1} : c \leq g\}$$

$\downarrow c_1$:

$$S_{k+1} = \{s \in S_k | s \leq c_1\}$$
$$G_{k+1} = \max\{c \in C | c \leq c_1, \exists g \in G_k : c \leq g, \exists s \in S_{k+1} : s \leq c\}$$

$\neg \uparrow c_1$:

$$S_{k+1} = \{s \in S_k | c_1\neg \leq s\}$$
$$G_{k+1} = \max\{c \in C | c_1\neg \leq c, \exists g \in G_k : c \leq g, \exists s \in S_{k+1} : s \leq c\}$$

$\neg \downarrow c_1$:

$$G_{k+1} = \{g \in G_k | c_1\neg \leq g\}$$
$$S_{k+1} = \min\{c \in C | c\neg \leq c_1, \exists s \in S_k : s \leq c, \exists g \in G_{k+1} : c \leq g\}$$

Figure 2.34 Finding the new G and S sets from the current G and S sets for each type of constraint.

parents and children is provided, the algorithm to update the boundary sets can be implemented efficiently.

Consider, for instance, constraint $\uparrow c_1$ in Figure 2.34 (the extension to the other cases is straightforward). The set G_k is simply updated by removing from it all elements that do not include c_1. The set S_k is updated generalizing all elements which do not include c_1 by the amount necessary to include it, and in such a way that they do not include any other class in S_{k+1} and they are included by some class in G_{k+1}. In particular, if a Hasse diagram of the concept lattice is available, the generalization is performed along the edges connecting each element to its more general elements.

It turns out that the worst-case time complexity of the updating of the admissible space is proportional to the square of the sizes of the boundary sets, which are bounded from above in many applications.

At this point we should emphasize that while the procedure described above allows us to find the intent of all concepts that satisfy the user constraints, the corresponding extent may still contain some *inadmissible* objects.

To explain, consider that the extent of a concept in an unconstrained lattice contains all objects that possess *at least* the attributes specified in the intent of that class, and that therefore the remaining attributes of some object may violate the constraints (i.e., the object also belongs to a forbidden class). We will see an example of this below.

Hence, the set of objects associated with a concept includes only those consistent with the constraints seen until that point. The algorithm introduced above can be easily completed in the following way to take into account the presence of inadmissible objects in valid concepts.

Since the only operation of the algorithm that is affected is the updating of the set S_k for constraints of type S_k and $\neg \downarrow c_1$, it is sufficient to check that in these cases the candidates of S_{k+1} have some admissible object.

The admissible objects of a concept are those which have the same description as the intent of either the concept or one of its admissible descendants; in practice, the set of admissible objects may be computed efficiently by marking all concepts between the two boundary sets (i.e., all admissible concepts) and collecting the objects of the relevant concepts.

As the overall approach that we have described is characterized by the presence of two boundary sets that delimit the admissible search space, it may be referred to as *bounding*. Clearly, the admissible search space after the application of one or more constraints will usually not form a sublattice of the original lattice.

To illustrate the bounding algorithm, consider pruning the concept lattice in Figure 1.5 with the two constraints $\downarrow f$ and $\neg \downarrow c$. In Figure 2.35, we show how the boundary sets are updated after each constraint is given as input to the system.

$\downarrow f$:
$S = \{$a b c d e f g h$\}$
$G = \{$e f$\}$

$\neg \downarrow c$:
$G = \{$e f$\}$
$S = \{$b e f h$\}$

Figure 2.35 Example pruning of the concept lattice in Figure 1.5 using the method described in Figure 2.34.

Determining the updated boundary sets in response to the second constraint requires further explanation. The concept with intent 'e f' in set G has only one admissible object associated with it (i.e., object 1), because the other two objects that belong to that concept in the unconstrained lattice also contain 'c' and thus must be removed. As for set S, the other possible generalizations of the concept with intent 'a b c d e f g h' (see the upper neighbours of the bottom element of the lattice in Figure 1.5) have been discarded either because they were not more specific than some element in G (i.e., 'a e g i', 'a c e g', 'd e h'), or because they did not comply with the $\neg \downarrow c$ constraint (i.e., 'c e f g', 'b c e f', 'c e f'). The updated set S contains just the concept 'b e f h' with its admissible object 1, the same as in set G.

In fact, the pruned search space has shrunk to three concepts (i.e., 'b e f h', 'b e f', 'e f') describing the same object, namely a bat (see Table 1.1). This is the only animal in the given context which has wings ($\downarrow f$) and does not have a beak ($\neg \downarrow c$).

So far we have analysed the problem of bounding the search space as new constraints are specified, until convergence to the target concept(s) is reached. In practice, the user may realize at some point during the search that some previously asserted constraint is too strong (e.g., when the version space becomes empty and the user is forced to abandon searching) or just wrong.

Although in certain cases it is possible to recover from this kind of inconsistency by expanding the original search space [32], the most obvious way to proceed is to retract some of the earlier constraints.

Unfortunately, the procedure that returns S_{k+1} and G_{k+1} given S_k and G_k cannot be inverted. Taking the constraint $\uparrow c_1$ as an example, it is easy to see that while in constraint assertion in order to find G_{k+1} it is sufficient to remove the elements that do not satisfy the last constraint from G_k, if we retract a constraint we have to consider both the elements that are above G_{k+1} and the elements that are incomparable to G_{k+1} in order to find G_k; furthermore, we cannot limit ourselves to testing such elements on the constraint being removed.

Due to the impossibility of using local searches to compute the result of constraint *retraction* from the current version space, one must compute from scratch the version space corresponding to the subset of constraints obtained by dropping the constraint to be retracted from the current constraint set. In practice, the computation of these new boundaries may be made more efficient considering that all the reasserted constraints can be grouped in two superconstraints – one including constraints of type $\uparrow c_1$ and $\neg \downarrow c_1$ and the other including constraints of type $\downarrow c_1$ and $\neg \uparrow c_1$ – and processed as if they were two normal incremental constraints.

2.5 Bibliographic notes

Equation (2.1) was developed by Godin, Saunders, and Gecsei [107]. In the same paper, different theoretical models for assigning attributes to objects are considered. Equation (2.2) was given by Carpineto and Romano [40]. Other theoretical results concerning enumeration of concepts and decision problems with constraints on the concept size have been found by Kuznetsov [137].

The Next Closure algorithm was first proposed by Ganter [91]. More recently, a detailed description and some refinements of this algorithm were provided by Kuznetsov and Obiedkov [139]. The version presented here is based on their work. Other interesting extensions to Next Closure include a post-processing step to generate the Hasse diagram [6] and a more efficient procedure to compute closures, based on an auxiliary tree [136]. A parallel version of Next Closure was recently presented by Fu and Mephu Nguifo [84].

Another algorithm for generating the set of concepts was proposed by Malgrange [153] and later refined by Chein [53]. It generates one level at a time starting from the object concepts; the next level is determined by systematically combining all pairs of concepts contained in the current level. Since the cardinality of each level may grow large, this algorithm is rather inefficient.

Turning to the methods for building both the set of concepts and the graph diagram, the Next Neighbours algorithm is based on the earlier work of Bordat [26]; the *FindLowerNeighbours* function used within it is based on more recent proposals by Carpineto and Romano [42] and Lindig [145]. The *Concepts Cover* algorithm is due to Nourine and Raynaud [172]; an incremental version of this algorithm is also available [173].

Regarding incremental algorithms, the Update by Cardinality algorithm was described by Godin and co-workers in [104] and [105]. The Update by Local Structure strategy was introduced by Carpineto and Romano under the GALOIS system ([34], [40]). Note that the version presented here is an improved and much more detailed one. Other incremental lattice-building algorithms that rely on the graph structure have been presented by Oosthuizen ([174], [175]), and by Valtchev and Missaoui [241].

Different strategies for building the full concept lattice have recently been explored, such as using the support function [230], similar to algorithms for mining association rules which will be discussed in Chapter 5, or applying the findings of graph theory [109] to a transformed representation of the input data.

The former approach has been implemented in the *Titanic* system. The concepts are generated in increasing order of cardinality (with respect to their intents), and the closure of each set is not computed with the doubleprime operator but by using the support of (or number of objects covered by) its immediate supersets. Furthermore, the closure of some sets does not need to be computed because it is equal to that produced by some other *key set*, or *minimal generator*.

Using the latter strategy, a particular (undirected) graph $G(CX)$ can be derived from a given context CX such that there is a correspondence between the concepts of CX and the *minimal separators* of $G(CX)$ [23]. At this point, a straightforward application of the algorithms developed to generate all the minimal separators (e.g., [133], [22]) would permit the construction of the full concept lattice in $O(|C||M+G|^2)$ time. Although this bound is no better than that of the best algorithms seen earlier, the existence of a link between the two fields might lead to new theoretical insights and more efficient methods.

It is also possible to construct a concept lattice as a sublattice of the product of two or more smaller lattices, similar to the nested line diagram visualization technique. This approach is termed *subdirect decomposition*. The implementation of methods for lattice construction based on subdirect decomposition is discussed in [86].

Relatively little work has been done on evaluating the performance of different lattice-building algorithms. Following an earlier analysis by Guénoche [114] and some experimental comparisons done by Godin *et al.* [105], a more systematic study has recently been presented by Kuznetsov and Obiedkov [139]. A bunch of algorithms is compared both theoretically and experimentally, and their worst-case time complexity is characterized using the additional notion of delay, i.e., the time interval between the generation of two consecutive structures.

Turning to the problem of dealing with a general change in the context table, an earlier algorithm for removing an object from a concept lattice was presented in [107]. A more complete approach to context change was then discussed in [36]. Our treatment is an elaboration of the latter work.

More recently, Valtchev and Missaoui [243] have addressed the problem of updating the lattice with a whole set of new objects rather than with a single object. Their approach was inspired by the subdirect decomposition paradigm: a separate concept lattice is built just for the new objects, and then the global lattice is computed by traversing the product of the original lattice and the new objects' lattice. The authors suggest that this method may be appropriate when the set of objects to be added is neither too small nor too large compared to the original set of objects; in addition, they make available a more efficient variant for the case when just one object at a time is added [241].

Automatic methods for graph drawing are discussed in [74] and [118]. Since their results may be unsatisfactory, as remarked in Section 2.3, it may also be useful to consider interactive methods. Freese [83] discusses the use of the force-directed placement technique to visualize concept lattices, with an interactive on-line prototype (http://www.math.hawaii.edu/ ralph/LatDraw). Another approach is described by Ganter and Wille [96]. It consists of assigning vectors to concepts (usually to meet irreducibles), laying out the corresponding lattice on the plane, and wandering through space of vector assignments until a regular diagram is found. An elaboration of the latter approach was presented by Cole [57], who investigated the use of *genetic algorithms* and force-directed placement to automatically find the vector assignments that optimize some objective function measuring the quality of the diagram. An alternative, and perhaps more effective, way of automatically finding a regular diagram is to prune a large proportion of the space by identifying diagrams that do not work in sublattices and then avoiding any elaboration of that diagram [58]. Sometimes, better line diagrams can be drawn by using an intermediate *geometrical* representation which helps understand the lattice-theoretic structure [96]. If the concept lattice is distributive, it can effectively be drawn through chain decomposition [260].

The textual indented representation used in Section 2.3.1 to visualize the hierarchy is probably neither the most compact nor the most visually pleasing tree layout. More dense and intriguing layouts have lately been proposed for representing large trees, such as cone trees [196], tree maps [129], information pyramids [11], or radial space-filling tree visualization [220]. On the other hand, a folder hierarchy is very simple, and it can be understood and manipulated more easily than the latter layouts. The tree-like representation can be found in some more recent lattice-based prototypes such as *HierMail* [62] or *Mail-Sleuth* (http://www.mail-sleuth.com).

It should also be noted that reducing a graph to a tree is not the only possible way of transforming the former into a simpler structure; examples of more complex transformations include using spanning trees [168] or layers [157]. A detailed account of the most common graph structures along with their associated visualizations is provided in [125]. Even the related problem of visualizing multiple overlapping hierarchies has recently received attention [111]. It is conceivable that similar techniques might be also useful for visualizing concept lattices, at least for some applications, although such attempts have not been reported so far in the literature.

Nested line diagrams have been developed and implemented mainly by Rudolf Wille and his co-workers ([255], [247], [248], [264]). Interesting refinements of the basic techniques have concerned the removal of visually redundant information, discussed earlier, and efficiency improvements. The latter issue has been addressed by Cole and Eklund [60], who describe a fast procedure for constructing the subcontext of interest (i.e., the context containing the documents with the same combination of selected attributes) from a one-valued, relatively large context in the medical domain. Nested line diagrams can also be drawn incrementally as new attributes are added to the subcontexts [113].

As mentioned in Section 2.3.2, the combination of multiple partial views has also been studied in the information visualization field (e.g., [195], [170]), where it is usually referred to as *linking* [98]. Among the research subjects which are more interesting from a concept lattice perspective, there is the question of what type of data sets and task are likely to benefit from multiple views [15] and the possibility of partially automating the generation of linked views [171]. A first step in the latter direction was taken by Cole and Eklund [60], who proposed the use of a thesaurus to drive the creation of useful conceptual scales prior to their combination in a nested line diagram. The combination of multiple representations or viewpoints has also actively been investigated in the information science field, with the of goal of facilitating the location of useful documents during a search or browse session (e.g., [193], [218], [108]).

Among the discrete solutions to the focus+context views, the DOI function was proposed by Furnas [88] as a practical solution to the fisheye visualization problem, although the effects observed when looking through a fisheye lens or magnifying glass could be reproduced much more closely by using non-trivial continuous spatial transformations such as those described in [132], [17] and [168].

The fisheye-view metaphor has been adopted for visualizing concept lattices in the ULYSSES prototype ([38], [39]). Simpler graphical interfaces based on offering just the neighbours have been suggested or implemented in several works, including [100], [106], [40] and [42].

Rapid advances in the field of graphical web interfaces may spur a renewed interest in the techniques for lattice visualization. In addition to exploring the use of alternative visual layouts, whether focused on the more complex inherent graph substructures mentioned above or on richer interactive or linking mechanisms (e.g., [205], [266], [112]), it would be useful to compare the relative merits and drawbacks of each visualization scheme for specific performance tasks.

Taking into account background knowledge has been proposed by several researchers, usually in the form of a partial ordering relation between the attributes of a (one-valued) context ([101], [35], [40], [215], [62]). These approaches model different situations but are conceptually similar. The expanded concept lattice is typically constructed from an explicitly expanded context using some standard lattice-building algorithm, with the notable exception of [40], where a method similar to that presented here was first proposed for one-valued contexts. For the case where a hierarchy is integrated into a concept lattice represented by a nested line diagram, we might be interested in computing the expanded lattice from the diagrams already drawn. This problem is discussed in [113].

Priss [190] discusses other possible ways in which a context and a subsumption hierarchy can be merged into an expanded context. She suggests that the user should be given the possibility of interactively combining concepts from multiple hierarchies, or hierarchy facets, using Boolean operators [191]. Direct manipulation of a description hierarchy is also allowed in the HierMail [62] or Mail-Sleuth system; the user can insert and remove descriptors and insert and remove ordering.

The background knowledge can also be modelled by considering implications between subsets of attributes (instead of implications between single attributes), or allowing 'descriptions' with a semilattice operation on them (rather than simple attributes with implications); see [94].

Using structured descriptors to influence concept formation has also been explored outside the concept data analysis community (e.g., [236]). Other forms of background knowledge have been proposed for specific clustering algorithms that may also have some value for concept lattices ([221], [233], [250]).

The bounding framework was first presented in [35] and was then reimplemented as one of the functionalities of the ULYSSES system ([38], [39]).

A more elaborate approach to interactively incorporating user knowledge into a concept lattice is described in [185]. The user may choose from representation languages characterized by different levels of granularity (e.g., by specifying optional attributes or restrictions on the set of values taken on by the attributes), and the system constructs the corresponding concept lattice. This process can be seen as using different levels of abstraction depending on the user needs.

The number of concepts present in a lattice can be also reduced by using syntactic criteria, such as the number of objects that support a concept or the cardinality of a concept's intent. This type of pruning is suggested, for instance, in ¡check¿[176] and [103]. A more elaborate approach [244] requires the construction of a particular bipartite graph embedding a concept lattice, termed an *entity attribute lattice*, from which certain types of concepts can be removed in such a way that the resulting structure is a complete sublattice.

Part II

Applications

3

Information Retrieval

Information retrieval is an iterative and interactive process which consists of submitting a query, seeing the ranked documents returned in response to the query, and submitting a new query, until the documents sought have been found or the search has been abandoned. Apart from the effectiveness of the query modification mechanisms, the result of the interaction between the user and the system crucially depends on the quality of the documents retrieved in response to each query.

In this chapter we discuss the use of concept lattices for these two fundamental information retrieval tasks, i.e., interactive query modification and automatic text ranking.

3.1 Query modification

Query refinement allows the user to recover from situations where the set of documents returned is too large, or, symmetrically, too small (or even empty).

Concept lattices can naturally be used for query refinement, because a query can be mapped onto a context-derived concept lattice and the user can exploit the lattice structure to modify the query in a gradual and incremental manner. Query refinement can further benefit from the incorporation of a thesaurus into the concept lattice. These issues are analysed in Sections 3.1.1 and 3.1.2.

Then in Section 3.1.3 we discuss how to generate index terms from documents. This allows us to deal with the common situation in which the document collection to be searched is not equipped with a set of index terms, thus significantly extending the scope of concept lattices for text retrieval.

3.1.1 Navigating around the query concept

In an information retrieval domain, the context is the usual term–document relation, i.e., the objects are the documents contained in the collection and the attributes are the index terms describing the documents.

Concept Data Analysis: Theory and Applications. Claudio Carpineto and Giovanni Romano
 ISBN: 0-470-85055-8

We briefly recapitulate the main characteristics of the resulting concept lattice, as analysed in previous chapters. Each concept (or node, as it will henceforth also be called), is a pair, composed of a subset of the index terms (i.e., the intent of the concept) and a subset of the documents (i.e., the extent of the concept); in each pair, the subset of terms contains just the terms shared by the subset of documents, and, similarly, the subset of documents contains just the documents sharing the subset of terms. The set of pairs is ordered by the standard set inclusion relation applied to the set of documents and terms that describe each pair. The partially ordered set is usually represented by a Hasse diagram, in which there is an edge between two nodes if and only if they are comparable and there is no other intermediate concept in the lattice. The ascending paths represent the subclass/superclass relation; the bottom concept is defined by the set of all terms and contains no documents, while the top concept contains all documents and is defined by their common terms (possibly none).

There are two fundamental observations that motivate the use of a concept lattice for query refinement. (i) Each node in the lattice can be seen as a query formed by a conjunction of terms (the concept intent) with the retrieved documents (the concept extent) (ii) Following edges departing upwards (downwards) from a query produces all *minimal conjunctive refinements* (*enlargements*) of the query with respect to that particular database, in the sense that there is no choice of term addition (deletion) which would produce an intermediate concept.

In practice, the extent of each concept is formed by documents that are likely to be relevant to the query corresponding to the concept intent because each such document contains all query terms (i.e., the extent can be seen as a subset of all documents that are relevant to that query). The lattice structure suggests which terms should be added to (or deleted from) those describing the intent to yield the minimal refinements (enlargements) of the query.

These properties can naturally be exploited to refine a user query based on the content of the collection being searched. One simple method is the following. Let the query be formed by a set of attributes $A \subseteq M$. After constructing the concept lattice of the document collection, the query is mapped onto the document lattice to find the concept corresponding to the query, i.e., the *query concept*. The query concept, determined by making use of the lattice structure, is the concept whose intent is equal to the query, if any, or the most general of the concepts with a larger intent than A. The query concept can also be characterized as the meet of the set of attribute concepts associated with each query attribute, i.e., $\bigwedge_{a \in A} \mu(a)$.

In particular, the query concept may coincide with the bottom element of the lattice for very specific queries. In this case, it may be more convenient to display as a query concept one or more best matching concepts rather than the bottom element of the lattice, which contains the entire set of index terms.

At this point the user is provided with a graphical interface which displays the query concept and its neighbours in the lattice, using some of the visualization schemes described in Section 2.3.

In fact, a different approach can be taken in which it is not necessary to build the concept lattice of the document collection as a preliminary step. One can determine the query concept using just the context, and then build a small portion of the concept lattice centred around the query concept. The extent of the query concept will be formed by all documents that contain the query terms (i.e., A'), and the intent will consist of all terms possessed by such documents (i.e., A''). Once the query concept (A', A'') has been determined, we can use the Nearest Neighbours algorithm described in Section 2.1.4 to build its immediate neighbours or more distantly related concepts.

This approach can be generalized to also handle Boolean queries (i.e., queries specified as Boolean expressions), for which the null-output, or, symmetrically, the information-overload, problem may be especially relevant. In this case, the extent of the query concept will contain all the documents that satisfy the Boolean query (if any), while the intent of the query concept is determined in the same way as for non-Boolean queries (meaning that the intent of the query concept will, in general, contain only *some* of the query terms).

To illustrate, we use an example screen from REFINER, a concept lattice-based system for refining Boolean queries. REFINER, described in [42], was implemented in Common Lisp on a Power Macintosh. Figure 3.1, taken from [42], was produced during an experiment in which a user was searching the documents of the CISI collection–a widely used bibliographical collection of 1460 information science documents described by a title and an abstract–relevant to the following question: 'Computerized information systems in fields related to chemistry'.

As seen in Figure 3.1, the query supplied by the user (i.e., 'computer AND chemistry') yielded the query concept with the intent {chemistry, computer, result, search}; i.e., all the documents in the database that had 'computer' and 'chemistry' also contained the terms 'result' and 'search'. Figure 3.1 also shows the list of documents associated with the encircled concept on the screen (and the text of one of them).

The main advantage of using a concept lattice for query refinement is that it allows a content-driven reformulation with a controlled amount of output. In the example given above, for instance, two query refinements are shown that would produce manageable outputs, whereas in the absence of this kind of information the user might be tempted to enlarge the initial query by deleting either 'computer' or 'chemistry', resulting in an output list containing hundreds of documents.

The display also reveals other terms in the database (e.g., 'chemical') that index documents of probable interest to the user. This is a form of text mining, which will be discussed in Chapter 4.

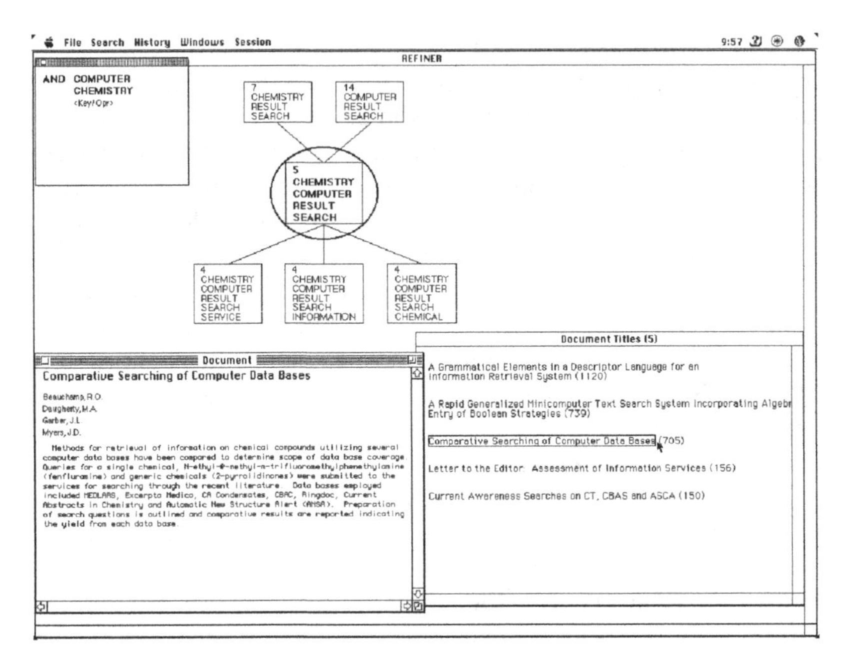

Figure 3.1 Refining Boolean queries with REFINER.

3.1.2 Thesaurus-enhanced navigation and querying

In a concept lattice, the proximity of two or more nodes is independent of the meaning of the terms describing the nodes, so that two semantically similar queries may be very distant within the lattice.

If a thesaurus over the terms contained in the document context is available, we can construct a concept lattice that incorporates the thesaurus information using the method described in Section 2.4.1.

In the resulting concept lattice, the ordering relation over the concepts is no longer independent of the possible ordering relation over the terms, because more general terms will index more general classes. Consequently, the user may find it easier to find the nodes containing semantically related queries, in that the region to be searched may reduce to a small subgraph of the whole lattice.

In addition to improving the usefulness of lattice browsing for query refinement, a thesaurus can be used to specify directly a more general or more specific query. It is sufficient to select one term in the current query concept and specialize/generalize it using the broader/narrower relationships contained in the thesaurus. We refer to this as *thesaurus climbing*. Query specification by thesaurus climbing lets the user make large jumps through the lattice driven by the semantics of terms.

To illustrate the use of thesaurus-enhanced concept lattices for query refinement, we consider again the simple document context in Table 2.2 along with the thesaurus in Figure 2.30. The corresponding concept lattice is shown in Figure 2.32. We describe a simple browsing session relating to this lattice, using the GALOIS-Browser user interface. GALOIS-Browser, described in [40], was implemented in Common Lisp and run on a Lisp machine. GALOIS-Browser displayed a focus node and its immediate neighbours and allowed three basic interaction modes: selection of a displayed node by direct graphical manipulation, query specification by term selection in a vocabulary window, and query refinement by term selection in a thesaurus window. The figures used here to illustrate the browsing session are taken from [40].

According to the visualization technique employed by GALOIS-Browser, each figure represents a screen displaying the current node (i.e., the black node) along with its parents and children. Simultaneously, the figure shows the action taken by the user to modify the current node, which results in the screen shown in the next figure.

In Figure 3.2 the system displays the most general query 'ARTIFICIAL-INTELLIGENCE, COMPUTER-APPLICATIONS, KNOWLEDGE-BASED-SYSTEMS' as the current query. Note that GALOIS-Browser displays for each intent only the terms that are not implied by any other term present in the intent itself. That is why the node with extent {1 2 3 4 5} is more specific than the node with extent {1 2 3 4 5 6} but has apparently fewer keywords; in fact, the term EXPERT-SYSTEMS is more specific than both COMPUTER-APPLICATIONS and KNOWLEDGE-BASED-SYSTEMS (see the keyword graph

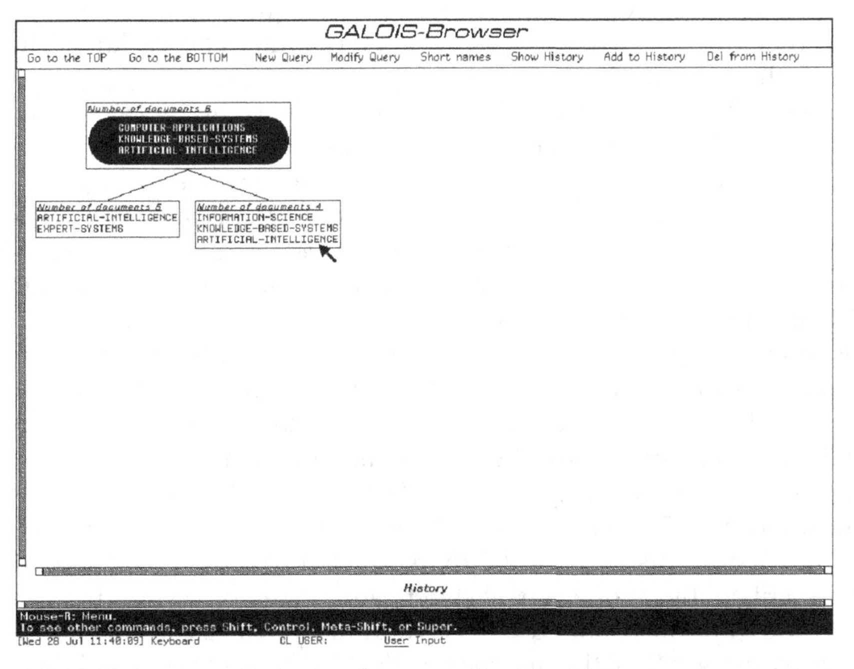

Figure 3.2 Initial screen displaying the current query 'ARTIFICIAL-INTELLIGENCE, COMPUTER-APPLICATIONS, KNOWLEDGE-BASED-SYSTEMS' and the action of the user to refine it.

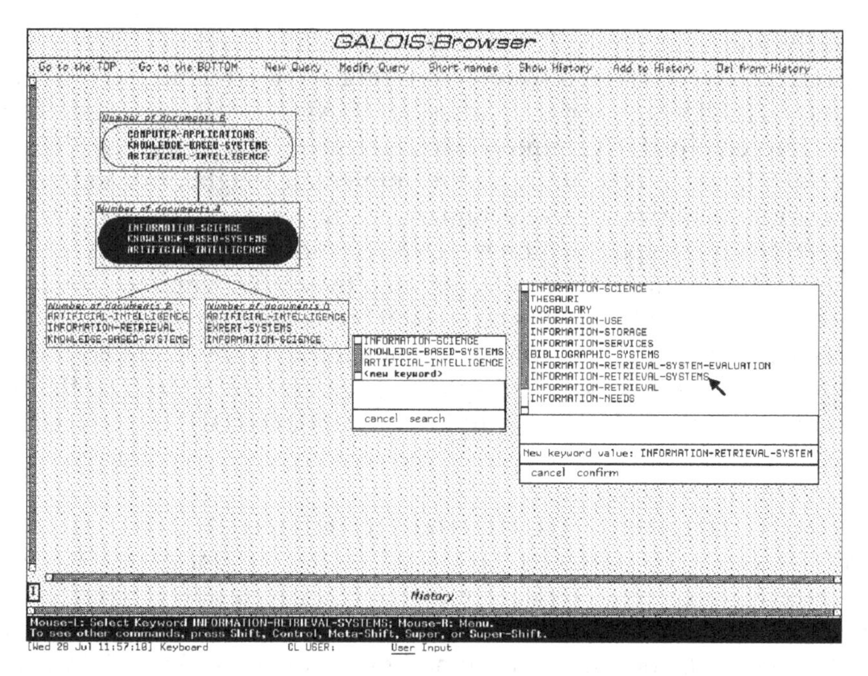

Figure 3.3 Specialization of the query by replacing the term *INFORMATION-SCIENCE* with the term *INFORMATION-RETRIEVAL-SYSTEMS.*

in Figure 2.30). The user selects the more specific query 'ARTIFICIAL-INTELLIGENCE, INFORMATION-SCIENCE, KNOWLEDGE-BASED-SYSTEMS' by pointing and clicking with the mouse on the right child of the current node.

In Figure 3.3 the user specializes the current query by replacing the term INFORMATION-SCIENCE with the term INFORMATION-RETRIEVAL-SYSTEMS. Since the query 'ARTIFICIAL-INTELLIGENCE, KNOWLEDGE-BASED-SYSTEMS, INFORMATION-RETRIEVAL-SYSTEMS' is not in the lattice, the system returns its most general specialization 'ARTIFICIAL-INTELLIGENCE, CATALOGUING, EXPERT-SYSTEMS, INFORMATION-RETRIEVAL-SYSTEMS', as shown in Figure 3.4. By clicking on such a node, the document associated with it (i.e., the fifth document of the example context) would be displayed.

It is useful to get an idea of how the use of a thesaurus affects the description of documents and the size of the concept lattice for real text collections. To this end, we report some data concerning a subset of the INSPEC computer-engineering collection, along with its thesaurus.

We queried INSPEC by questioning 'artificial intelligence', and selected the most recent 1555 elements out of some 10 000 documents retrieved, of which the documents in Table 2.2 represent a tiny sample. The documents were described by a title, an abstract, and a set of terms. In order to deal with a controlled and compact vocabulary, we used only the terms labelled as *preferred*. With this choice, there were 926 distinct keywords, with an average of 5.10 terms per document. The corresponding concept

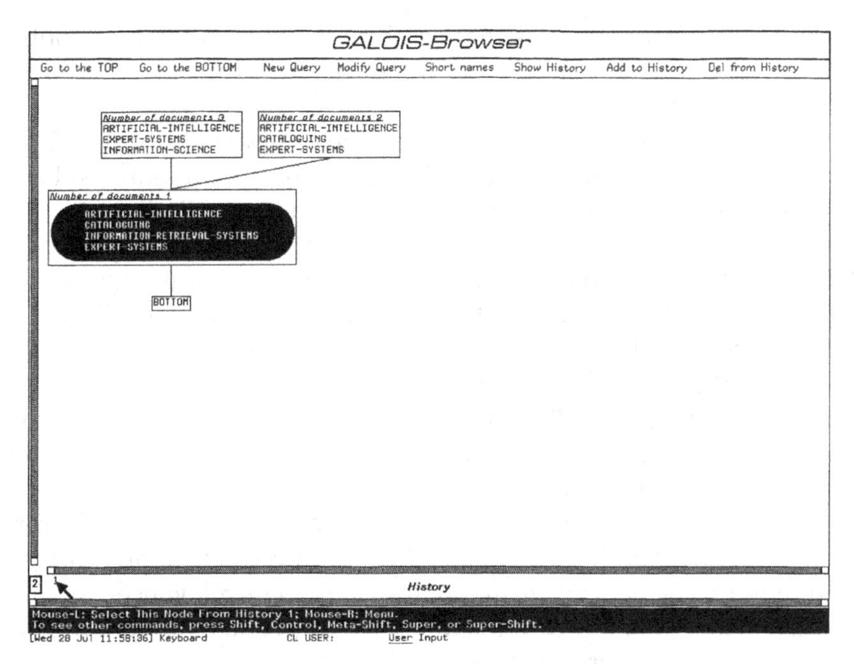

Figure 3.4 Result of the query 'ARTIFICIAL-INTELLIGENCE, KNOWLEDGE-BASED-SYSTEMS, INFORMATION-RETRIEVAL-SYSTEMS'.

lattice consisted of 3763 nodes with an average of 2.88 parents per node and path lengths ranging from 3 to 11 edges from the lattice's top to bottom node.

Building the thesaurus-enhanced concept lattice required some more pre-processing. From the broader/narrower relation among preferred terms given in the 1991 INSPEC thesaurus, we extracted the subgraph containing all terms equal to or broader than the 926 terms used to describe the documents. The resulting graph, of which Figure 2.30 represents a small portion, contained 1136 terms. With this enlarged set of keywords each document was described by 10.15 terms on average. The thesaurus-enhanced lattice contained 8769 nodes, with an average of 3.11 parents per node and a depth ranging from 2 to 15 edges.

Thus, the incorporation of the thesaurus into the INSPEC lattice resulted in a considerable increase of the number of nodes (from 3763 to 8769) and caused the lattice to become deeper, without significantly affecting the mean number of parents (or children).

It is also interesting to evaluate whether enriching a concept lattice with a thesaurus really facilitates query refinement, with an improvement of retrieval performance. The evaluation issue is of central importance to any interactive information retrieval system, including those based on concept lattices. However, it is not easy to conduct this type of experiment due to the presence of a number of factors that can bias the results in an uncontrolled manner (background of subjects, user interface, test protocol, environmental characteristics, etc.).

We performed an experiment to evaluate the comparative retrieval performance of using a document lattice with (and without) a thesaurus. Here we briefly summarize the experimental design and the main findings.

We asked some external users to perform a subject searching task on the INSPEC subcollection illustrated above using two systems, i.e., GALOIS-Browser applied to the concept lattice of the INSPEC subcollection and GALOIS-Browser applied to the concept lattice of the INSPEC subcollection augmented by the INSPEC thesaurus.

Because the INSPEC subcollection did not have a set of test queries with their associated relevant documents, as for instance with the currently available TREC (Text REtrieval Conference, http://trec.nist.gov) and CLEF (Cross-Language Evaluation Forum, http://clef.iei.pi.cnr.it:2002) test collections, it was necessary to create them manually. We produced 20 queries with mixed difficulty, and for each query we manually assessed its relevance judgements, i.e., the associated set of relevant documents. The average number of relevant documents for the 20 queries was 30.5.

The users had a computer science background, but had little knowledge of the document domain and did not know the GALOIS-Browser system. We provided them with a tutorial session on a small training database until they could use the system effectively.

For assigning the queries to the two retrieval methods (i.e., unenhanced and enhanced concept lattice) we used a repeated-measures design, in which each user searched each query using each method. Sequence effects were reduced by varying the order of the two methods over the query set.

During each search the user, who was not asked to finish within a certain time period, could see the abstracts of the documents associated with the nodes visited. The documents judged to be relevant by the user, as well as those scanned during the search, were noted as retrieved. The user system sessions were fully recorded. For each search we considered four measures: recall, precision, number of nodes visited, and search time (i.e., the time taken by the user to perform his or her task).

The method with enhanced lattice performed better than the method with unenhanced lattice. We found that the method with thesaurus achieved better results than the method without thesaurus for recall and search time and had comparable results for precision and number of nodes visited. This experimental evidence further supports the view that a thesaurus-enhanced lattice can be navigated and queried faster and more effectively than an unenhanced lattice.

So far, we have implicitly assumed that a set of index terms describing each document is available. This was actually the case for the INSPEC collection, whereas for the CISI collection we automatically extracted a reduced set of terms describing each document from the full-text document descriptions.

In fact, in most information retrieval applications the set of documents comes without index terms and their manual generation is often impractical or even unfeasible (think of large text databases that change frequently over time).

Full-text retrieval is easily handled by most statistical information retrieval systems, but it is not practical for concept lattice-based applications, for which we need to generate a restricted set of index terms. This issue is addressed in the next subsection.

3.1.3 Automatic generation of index terms

Automatic indexing has been studied in information retrieval. To automatically extract a set of index terms describing each document, the following steps can be followed.

1. *Text segmentation*. The individual words occurring in a text collection are extracted, ignoring punctuation and case.

2. *Word stemming*. Each word is reduced to word-stem form. This may be done with some rule-based stemmer or using a lexical knowledge base.

3. *Stop wording*. A stop list is used to delete from the texts the (root) words that are insufficiently specific to represent content, such as *the*, *of*, *this*, *on*, and some verbs, such as *have*, *can*, *indicate*.

4. *Word weighting*. Word weighting is at the heart of any document ranking system. In our setting, this step is necessary to perform word selection, described in step 5; it may also be useful for concept-lattice based text ranking, described in the next section.

The typical approach to word weighting is as follows. For each document and for each term, a measure of the usefulness of that term in that document is derived. The goal is to identify words that characterize the document to which they are assigned, while also discriminating it from the remainder of the collection. This has long been modelled by the well-known 'term frequency–inverse document frequency' (tf–idf) scheme. Term frequency is given by the ratio of the number of times a term occurs in a document to the total number of terms in that document. Inverse document frequency is the total number of documents in a collection over the number of documents in which the term occurs.

The two typical assumptions of the tf–idf scheme–namely that multiple appearances of a term in a document are more important than single appearance (tf) and that rare terms are more important than frequent terms (idf)–are

BM25:

$$w_{t,d} = \frac{(k_1 + 1) \cdot f_{t,d}}{k_1 \cdot \left[(1 - b) + b \dfrac{l_d}{\mathrm{av}(l_d)}\right] + f_{t,d}}$$

SLM:

$$w_{t,d} = \log_2 \frac{f_{t,d} + \mu\lambda_t}{l_d + \mu} - \log_2 \frac{\mu}{l_d + \mu} - \log_2 \lambda_t + \cdot \log_2 \frac{\mu}{l_d + \mu}$$

DFR:

$$w_{t,d} = \left(\log_2(1 + \lambda_t) + f^*_{t,d} \cdot \log_2 \frac{1 + \lambda_t}{\lambda_t}\right) \cdot \frac{f_t + 1}{n_t \cdot (f^*_{t,d} + 1)}$$

with $f^*_{t,d} = f_t \cdot \log_2 \left(1 + \frac{c \cdot \mathrm{av}(l_d)}{l_d}\right)$

$w_{t,d}$	the weight of term t in document d
f_t	the number of occurrences of term t in the collection
$f_{t,d}$	the number of occurrences of term t in document d
n_t	the number of documents in which term t occurs
D	the number of documents in the collection
T	the number of terms in the collection
λ_t	the ratio between f_t and T
l_d	the length of document d
$\mathrm{av}(l_d)$	the average length of documents in the collection

Figure 3.5 Effective term weighting schemes.

usually extended through a third length normalization assumption, which states that for the same quantity of term matching, long documents are less important than short documents.

This three-component framework has been implemented using several approaches, the best known of which is probably the vector space model [204]. In Figure 3.5 we show three more recent weighting functions: BM25 [200], statistical language modelling (SLM) [278], and deviation from randomness (DFR) [10]. The formulae in Figure 3.5 have been adapted to the Task at hand, in which the documents are not scored against a user query, and contain some collection-dependent parameters. Typical values are as follows: for BM25, $k_1 = 1.2$, $k_3 = 1000$, $b = 0.75$; for SLM, $\mu = 1000$; for DFR, $c = 2$. These functions have been shown to perform very well on large, heterogeneous test collections, such as those used at TREC and CLEF. In particular, BM25 has been used by most participants in TREC and CLEF in recent years.

For semistructured or Web documents, text-based indexing might be complemented with other techniques that take advantage of additional sources of knowledge, such as document fields, incoming or outgoing links, anchor texts, and URL structure.

5. *Word selection*. This last step is not necessary for information retrieval systems performing full-text indexing, but it is important for concept lattice-based systems for facilitating the subsequent process of lattice construction.

Word selection can be performed using some heuristic threshold with which to select the index terms with the highest weights. For instance, one can use as selection criterion the mean of weights in the document or the value corresponding to one standard deviation above the mean.

Clearly, the index terms selected ultimately control the retrieval performance of the overall system. Note that the extraction of a good set of index terms is one of the most difficult steps when producing a useful document lattice, because an optimal strategy should consider the characteristics of the collection at hand and the specific queries that will be submitted for refinement.

3.2 Document ranking

Not only can the concept lattice be used for interactive query refinement but also for fully automatic text ranking. In the latter case, the lattice structure serves to compute a conceptual distance between the query and each document, favouring the retrieval of relevant documents that do not necessarily match the query terms.

The application of concept lattices to text ranking is discussed in this section. We first review the vocabulary problem, faced by most conventional information retrieval systems, then we present concept lattice-based text ranking, and finally we discuss the scalability issue, namely how well the technique will work on a large data collection.

3.2.1 The vocabulary problem

The typical setting of a text ranking application consists of a user submitting a query to the system, usually in free-text natural language, and the system returning a list of possibly relevant documents in ranked order. Most ranking systems perform best-match retrieval, i.e., they build an internal representation of query and documents in the form of weighted term vectors and then score the query vector against the document vectors to produce a ranked document list.

Best-match ranking (BMR) systems are highly efficient and perform well in many situations; however, they are limited by their inability to deal with word mismatch. When translating an information requirement into a query for a document retrieval system, a user must convert concepts involved in his/her requirement into query terms which will not necessarily match the terms used by the authors to describe the same concepts in their documents. This is the well-known vocabulary problem, two specific important aspects of which are polysemy (the same word describing different things) and synonymy (different words to describe the same thing).

The severity of the vocabulary problem is exacerbated when the queries are very short and the collection being searched is large and heterogeneous in content, as with Web-based retrieval. The paucity of query terms increases the difficulty of handling synonymy, because there is less chance of some important word co-occurring in the query and in the relevant documents. The large number of heterogeneous documents being searched makes the effects of polysemy more evident, as there is a greater chance of an ambiguous word co-occurring in the query and in the non-relevant documents. As a result, the system may fail to retrieve the relevant documents while retrieving many irrelevant documents.

A common solution to the vocabulary problem is to exploit the relationships in content which exist between the documents in the collection when deciding which documents are to be retrieved in response to a query. The best-known approach is cluster-based ranking, where a query is ranked not against individual documents but against a hierarchically grouped set of document clusters.

Hierarchical clustering-based ranking (HCR) methods take as input a matrix of document-to-document distances, based on a similarity function, and iteratively merge the most similar pair of distinct clusters, using a clustering strategy (e.g., single link, complete link, group average, Ward's method), until there is only one cluster. Once the clustering hierarchy has been built, an incoming query is ranked against neighbourhoods of this structure, using a search strategy (top-down, bottom-up, or optimal search) and a query-cluster similarity function.

The result of HCR is a partially ordered set of documents, because the documents in each cluster are equally ranked; however, a totally ordered list can be easily obtained from it by individually ranking the documents in each cluster against the query.

HCR is more robust than BMR with respect to word mismatch because it takes into account both the interdocument similarity and the similarity between a query and the individual documents. However, HCR too has some important limitations. It has a relatively small number of free parameters (essentially, n clusters for n documents), it is unable to perform multiple or crossed classifications, and it relies on heuristic strategies.

The last issue is, perhaps, the most important, because it affects the correctness of the results. In particular, HCR may fail to discriminate between documents that have manifestly different degrees of relevance for a certain query,

Table 3.1 A simple document-term relation

	T1	T2	T3	T4
D1	×	×		
D2		×	×	
D3			×	×

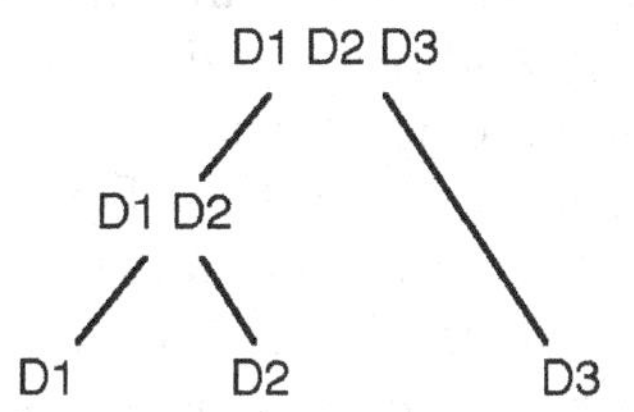

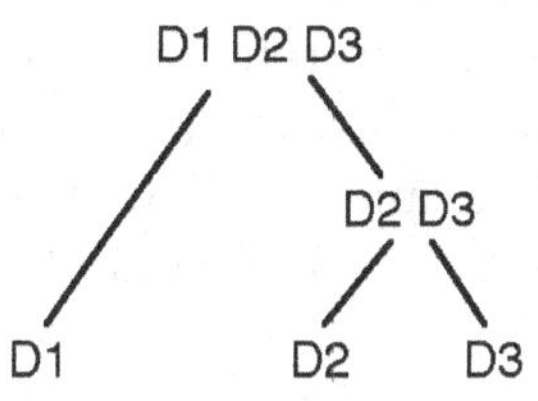

Figure 3.6 Two equally plausible cluster hierarchies derived from the context in Table 3.1.

even for very simple databases and in the presence of typical assumptions and choices.

To illustrate, we will use an example taken from [44]. Consider the simple document–term relation described in Table 3.1. Clearly, BMR methods would always fail to discriminate between documents that have no terms in common with a query. For instance, if the query is equal to 'T1', documents D2 and D3 would be equally ranked, whereas D2 appears to be more relevant than D3 for the given query. If we use HCR, clusters {D1, D2} and {D2, D3} are equivalent regardless of the chosen interdocument similarity function; thus, we may have two possible cluster hierarchies for the set of documents at hand (see Figure 3.6).

For query 'T1', the left hierarchy would produce a correct document ranking, because the best-matching cluster would be {D1}, followed by cluster {*D*1, D2}. The right hierarchy, however, would rank cluster {D1, D2, D3} right after cluster {D1}, which would cause documents D2 and D3 to be equally ranked. This behaviour will be observed for any choice of the query-cluster similarity function.

The limitations of HCR can be overcome by using the concept lattice of the document collection as the underlying clustering structure and computing the similarity between the query and the documents using the structure itself rather than a heuristic metric. This approach is described in the next section.

3.2.2 Concept lattice-based ranking

The first observation for motivating concept lattice-based ranking (CLR) is that the matching between a query and a document can be described in terms of a sequence of operations that transforms one into the other. In particular, the basic operations to transform the term vector representing a query into the term vector representing a document can be seen as a form of term addition and term deletion, while the length of the sequence that accomplishes the desired transformation is a measure of the similarity between the two vectors.

Note that one apparently straightforward way to implement this idea is simply to count the terms that are not shared by the query vector and the document vector (or use some related similarity measure such as the Dice or Jaccard coefficient). This method, however, would amount to a simplistic form of best-match retrieval, with very limited ranking abilities. Neither would it be able, in general, to discriminate between non-matching documents. In the simple example considered in Table 3.1, for instance, document D2 would not be ranked ahead of D3 for query 'T1', because the two documents would get the same distance score. Thus, viewing the distance between a query and a document as a transformation process between the corresponding representations is not sufficient in itself.

The second intuition behind this approach is that the query–document transformation should be driven by a conceptual representation of the database being searched, in which both the queries and the documents can be merged.

A document lattice enjoys several useful properties for acting as such an intermediate representation: (i) the concepts have a natural interpretation from the point of view of characterizing the set of queries that can be satisfied by the collection; (ii) a user query can be mapped onto a query concept (see Section 3.1.1); (iii) the concept ordering tells us how to gradually pass from one query to another; and (iv) the set of concepts in the lattice is sufficiently numerous to ensure a rich representation while being still tractable.

Furthermore, just as a query concept can be assigned to each given query, a *document concept* can be assigned to each document in the collection. The document concept of a document d is given by (d'', d'), where d' is the document intent (i.e., the set of attributes (terms) describing d). This is exactly the definition of an object concept, considering that in this case the objects are documents (see Section 1.2).

At this point, the problem of computing a sequence of operations that transforms a query into a document can be cast as a breadth-first search through the partially ordered space of admissible queries, as defined by the document lattice. The initial state is the query concept, while the successor states are specified by the *nearest neighbour relation* (denoted $\succ\!\prec$). The search ends as soon as the document concept is reached. A document score is given by the length of the shortest path linking the query concept to the document concept, while the final ranked document list is obtained by arranging the documents in increasing order of score. We now provide a more precise description of the method just sketched.

Let D be the set of documents, T the set of terms describing the documents, and I the document $\times$ term matrix. Consider the ordered set $(\mathcal{C}(D,T,I); \succ\prec)$ formed by the set of concepts of the context (D,T,I) along with the nearest neighbour relation, i.e., for $x, y \in \mathcal{C}(D,T,I), x \succ\prec y$ if $x \succ y$ or $y \succ x$. Define the *distance* between concepts x and y as the least $n \in \mathcal{N}$ for which the following condition holds:

$$\exists z_0, z_1, \ldots, z_n \in (\mathcal{C}(D,T,I); \succ\prec) \text{ such that } x$$
$$= z_0 \succ\prec z_1 \succ\prec \ldots \succ\prec z_n = y.$$

For each query q and $d_1, d_2 \in D$, d_1 is ranked ahead of d_2 with respect to q if and only if the distance between the query concept (q'', q') and the document concept (d_1'', d_1') is less than the distance between (q'', q') and the document concept (d_2'', d_2').

It follows that if a document d_1 is ranked ahead of a document d_2 for a user query q, then the set of terms contained in d_1 can be derived from the set of terms contained in q by a smaller number of admissible minimal transformations than those necessary to derive d_2.

From a computational point of view, the query is merged into the document lattice and each document is ranked according the length of the shortest path linking the query to the document concept. We can think of the sets containing equally ranked documents as concentric *rings* around the query node: the longer the radius, the lower the document score (of the associated document concepts). Of course, this is a partially ordered retrieval output, because the documents that are equally distant from the query concept have the same score.

Thus, a natural completion of this method consists of individually ranking the documents in each ring against the query. Using BMR, the similarity between the query q and each document d can be computed as the product of a document-based term weight by a query-based term weight, i.e., $\text{sim}(q, d) = \sum_{t \in q \wedge d} w_{t,d} \cdot w_{t,q}$. For the document-based term weight, we can use one of the expressions given in Figure 3.5; for the query-based term weight it is sufficient to consider the number of occurrences of the term in the query.

Alternative ways of ranking the documents contained in each ring are also possible. For instance, one could discriminate between two concepts that are connected with the query through paths of the same length by using the number of distinct paths of that length that link the query to either concept. Such an approach would represent a more natural refinement of our main ranking criterion, because it would be based on a measure of the degree of multiple structural connections between the concepts; however, this would be much more inefficient than BMR.

It should also be noted that, in practice, it is convenient to omit both the top and the bottom elements of the lattice in the graph used to compute the partially ordered set of documents, because these two concepts cannot be seen as proper queries with associated documents. The only exception is when the top element of the lattice is described by a non-empty set of terms (i.e., all documents have those terms), or, dually, when the bottom element of the lattice contains a non-empty set of documents (i.e., one or more documents have all terms).

Consider again the example context given in Table 3.1. Figure 3.7 shows the corresponding concept lattice, in which the top and the bottom elements have been omitted. For query 'T1', using CLR the first document retrieved would be D1, because there is a link connecting the query to it, followed by document D2, through the path T1–T1T2–T2–T2T3, and by document D3, through the path T1–T1T2–T2–T2T3–T3–T3T4, which reflects the actual relevance of the three documents to the given query.

We now present a more elaborate illustrative example, taken from [44]. We refer to the simple bibliographic database shown in Table 3.2. All documents are about computerized decision support systems, with two broad application domains: economics and the environment. Suppose we are interested in the

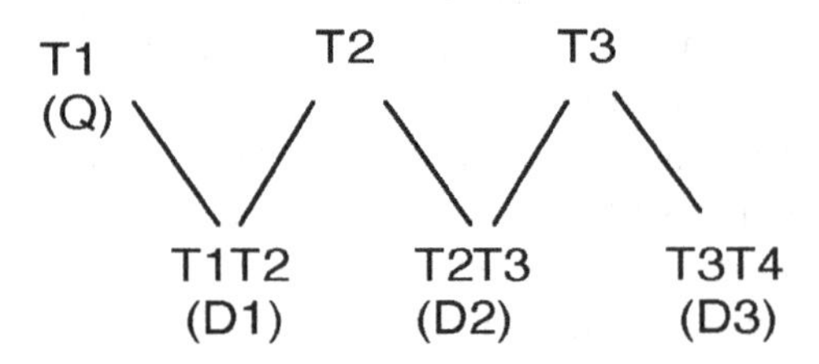

Figure 3.7 The concept lattice of the context in Table 3.1, without the top and bottom elements.

Table 3.2 A simple bibliographic context containing seven documents described by eight index terms

	D1	D2	D3	D4	D5	D6	D7
Neural-Network-Systems	×	×	×				×
Knowledge-Based-Systems				×	×	×	×
Credit				×	×		×
Finance	×			×			×
Account	×		×				
Bank	×	×	×			×	
River		×					
Waters		×					

former class of documents, and let us assume that the database is queried by questioning 'NEURAL-NETWORK-SYSTEMS, FINANCE'.

In Figure 3.8 we show the concept lattice built from the documents at hand, omitting the top and the bottom elements and denoting NEURAL-NETWORK-SYSTEMS and KNOWLEDGE-BASED-SYSTEMS by NNS and KBS, respectively. To visualize the diagram, we use a focus+context representation centred around the query concept of the query 'NNS, FINANCE', i.e., the concept with intent {NNS, FINANCE}. The concepts are displayed in decreasing levels of detail (with respect to the font size) and at increasing graphical distance, depending on their distance from the focus.

The labelling of each concept is such that the intents are shown in full for every concept, while minimal labelling is used for the extent (i.e., each document appears only once, in the corresponding document concept). For each concept, the distance from the query concept is shown in the upper right-hand corner, thus identifying the ring to which the concept belongs.

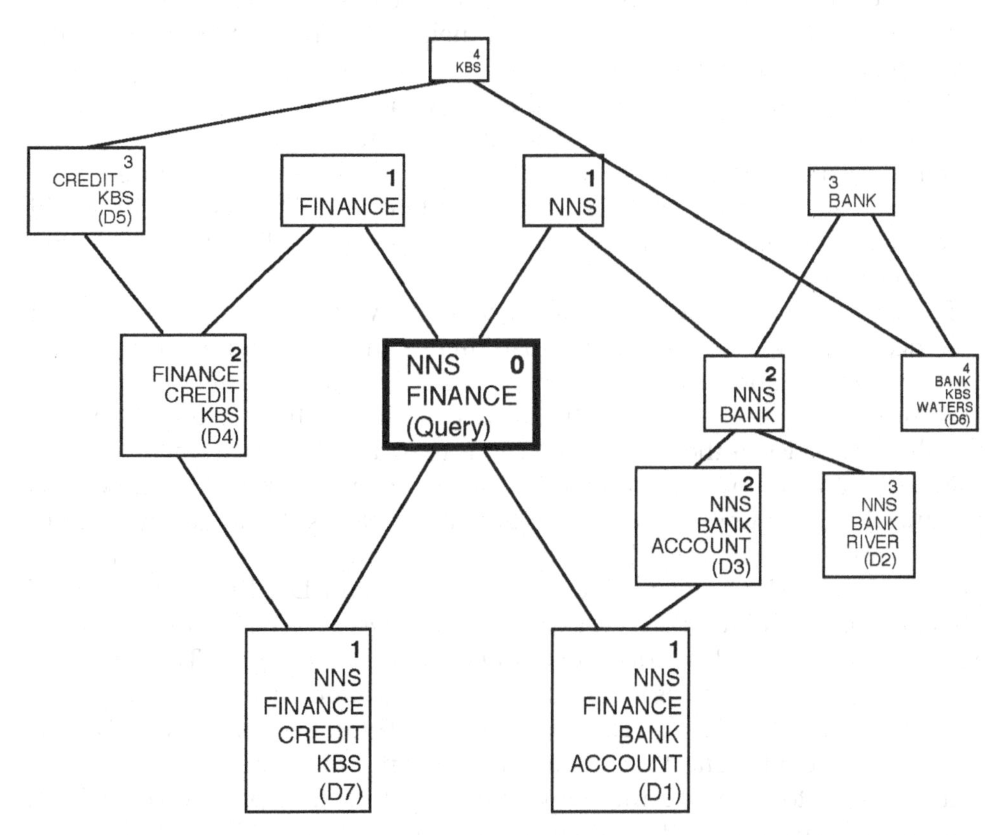

Figure 3.8 Concept lattice-based ranking of the documents in Table 3.2 for the query 'NEURAL-NETWORK-SYSTEMS, FINANCE'. Four concentric rings are visible, denoted by the number in the top right of each concept box.

Thanks to the focus + context visualization technique, four concentric rings around the query concept are visible in Figure 3.8.

We have seen that, in general, the completeness constraint on which concept lattices are based limits the number of admissible concepts by favouring maximally specific descriptions. In an information retrieval setting, this is equivalent to assuming that if one term always appears jointly with other terms, the single terms do not refer to distinct entities while their tuple conveys a useful meaning. In a sense, every complete set of co-occurrences determines a semantic context specific to the collection at hand, and the closeness between two semantic contexts is completely determined by the distribution of terms in the collection's documents.

One consequence is that in the lattice there may well be two near concepts that differ by a larger number of terms than more distant concepts do. For instance, the query concept and the concept with extent D7 are connected by a single-link path although they differ by two terms (i.e., CREDIT and KBS), because creating an intermediate concept between the query concept and D7 with either CREDIT or KBS would be of no help in discriminating between the given set of documents.

At the same time, it may happen that the same two concepts become more distant, or closer, as the set of documents in the collection changes. For instance, as a consequence of the introduction of a new document described by NNS-FINANCE-CREDIT, a new concept with intent NNS-Finance-Credit would be added to the lattice between the query concept and D7, thus increasing their distance.

Thus, this kind of distance is a context-sensitive measure, similar to most distance measures based on some clustering method. By contrast, this important feature cannot be easily incorporated into more traditional statistical measures where the distance between two representations is based only on the characteristics of those representations (e.g., Euclidean distance).

Now let us see how the ranked list for the given query and documents is produced. The (intent of the) concepts that are closest to the query concept (ring 1) are: {FINANCE}, {NNS}, {NNS, FINANCE, CREDIT, KBS}, and {NNS, FINANCE, BANK, ACCOUNT}. The relevant documents are D1 and D7. Ring 2 consists of the concepts with intent {NNS, BANK}, {FINANCE, CREDIT, KBS}, and {NNS, BANK, ACCOUNT}, which yield the documents D3 and D4, and so on. The complete ranked list of documents is the following (with distances from the query node in parentheses): D1 (1), D7 (1), D3 (2), D4 (2), D2 (3), D5 (3), and D6 (4). This result does seem to reflect the document relevance to the given query.

It is useful to compare the above ranking with that produced by BMR methods. We would get the following ranked list: D1 and D7, with the same score, followed by D2 and D3, with the same score, then D4, and then D5 and D6, with the same score. This result presents several inversions with respect to the ideal ordering, involving not only pairs of documents that did not match the query (D5 and D6), but also pairs where only one document matched the query (D2 and D5), or both documents matched the query (D2 and D3).

The last case is interesting. Documents D2 and D3 have the same term in common with the query 'NNS' and both contain the ambiguous word 'Bank'. Using the lattice, the concept intent {NNS, BANK} can be equivalently specialized with 'RIVER' and with 'ACCOUNT' (see Figure 3.8); however, the concept intent {NNS, BANK, ACCOUNT} may be reached from the given query through a shorter alternative path, which reflects the higher overall structural similarity of the latter concept with the query.

The computational time complexity of CLR is essentially determined by the complexity of lattice construction, because the computation of the ranked list of documents from the query and the lattice requires at most one pass over the lattice itself. The detailed working of the algorithm is as follows.

Starting at the query concept, a bidirectional breadth-first search through the lattice is performed until all nodes have been reached, without generating the concepts that have already been encountered. Then, if there are some disconnected concepts left, they are added at the end of the ranked list that has been produced so far. Disconnected concepts may arise as a consequence of the removal of the top and the bottom elements of the lattice; in practice, however, such disconnected concepts should be rare, and the corresponding documents, if any, are likely to be of no interest to the given query.

Each step of the breadth-first search returns a set of concepts that are equally distant from the query, with an associated set of documents. Thus, the position of each document in the output ranked list is computed in parallel while visiting the concept lattice.

3.2.3 Scalability

Given the computational requirements of concept lattice construction, text ranking based on concept lattices is suitable for small to medium-size document collections (up to tens of thousands of documents), whereas it cannot be applied, at least in the form described, to large databases containing millions of documents or more. Here we briefly discuss how this approach can be modified to scale up to large collections.

One approximate solution is to build a partial lattice centred around the query concept using the Nearest Neighbours algorithm presented in Section 2.1.4. If it is not necessary to generate all rings (e.g., when we are only interested in the most relevant documents), this method may be more efficient than full lattice construction.

Since the number of neighbours of each concept is usually proportional to the size of the concept intent, irrespective of the number of documents in the collection, finding all the rings at a distance of at most n from the query concept will take, for a query with q terms, at most $1 + q + q^2 + \cdots + q^{n-1}$ invocations of the Nearest Neighbours algorithm. On the other hand, the first rings usually account for a small but significant fraction of the total number of relevant documents.

The scalability issue can also be addressed by combining CLR and BMR, which seem to complement each other very well. BMR is more efficient and may have better effectiveness when considering the whole set of documents, while CLR may be more effective for the first documents retrieved and can discriminate between non-matching documents. A combined strategy keeps the strengths of the two methods while avoiding their main weaknesses.

One simple integration strategy is to have BMR rank an entire collection first, and then to use CLR to refine the ranking of the best-matching documents returned by BMR. The anticipated advantages of such an integrated approach are that it can be applied to large-scale collections and should feature a better retrieval performance than BMR on the first documents retrieved, which are of greater interest in many practical applications.

3.3 Bibliographic notes

Some lattice representations were used in early information retrieval [216], and even more recently [219], for refining queries containing Boolean operators. However, as these approaches typically rely on a Boolean lattice formalization of the query, the number of refinements proposed may grow too large even for a limited number of terms and most of the refinements will, in any event, be semantically meaningless to the user.

Methods for supporting Boolean query refinement that are, in nature, more heuristic have also been presented, such as weakening an AND clause by removing some terms (called *quorum-level searches* [55]), or using a set of rules for modifying the constituents of a Boolean expression (including terms, operators, and precedence) when the user faces an adverse situation [27].

Query refinement was one of the earliest applications of concept lattices to the information retrieval field [100]. The REFINER system is described in [42], along with an experimental evaluation showing that the effectiveness of information retrieval using REFINER is better than unrefined, conventional Boolean retrieval.

The mapping of a query onto the lattice can be done by choosing the most general concept that contains all the query terms, similar to REFINER, or with some weaker criteria if such a concept coincides with the bottom element of the lattice [244].

Once the query has been mapped onto a concept, the user may choose one of the neighbours of that concept, as in REFINER, or select one term from a list containing all the terms that are contained in the descendants of that concept [144]. In the latter case, the concept lattice must be fully built.

The GALOIS-Browser system was presented in [40], along with a more detailed experimental evaluation than reported here. That the integration of a thesaurus into a cluster-based (or interactive retrieval) method can improve recall without sacrificing precision was also advocated in [87] and [272].

Interactive information retrieval systems based on concept lattices have been developed for a large variety of textual and semistructured data, such as animation films [106], the INSPEC computer-engineering collection ([40], [39]), medical documentation ([59], [60]), the CISI information science collection [39], software documentation ([144], [191]), classified advertisements for real estate [61], file systems [79], e-mail messages ([63], [62]), bioinformatics data [78], and the CLEF news collection [54].

The evaluation of the retrieval performance of an interactive information retrieval system is made difficult by a combination of methodological, technological, organizational and economic issues. The techniques developed in the field of human–computer interaction ([212], [189]) and experimental psychology [201] can help design more reliable tests involving human subjects. Two good examples of evaluation studies of interactive information retrieval are [218] and [21]. Most of the interactive information retrieval systems based on concept lattices have not been experimentally evaluated, although there are some significant exceptions ([101], [40], [39], [42]).

Turning to the generation of index terms, a classic blueprint for automatic text indexing was put forward by Salton [203].

The most popular rule-based stemmer is based on Porter's algorithm [188]; several versions of this algorithm have been made available on the Snowball website (http://snowball.tartarus.org) for different European languages. More sophisticated stemming approaches require the use of linguistic resources; a good example of a large morphological lexicon that contains the standard inflections for nouns, verbs and adjectives is [130].

Several *stop lists* are available. The stop list included in the CACM data set, for instance, contains 428 common function words. Stop lists for languages other than English are also available [207].

The BM25 term weighting formula [200] is the last refinement of one of the best-known probabilistic information retrieval models, first proposed by Robertson and Sparck Jones [198] and subsequently improved [199]. The statistical language modelling approach has been proposed in several papers, with many variants (e.g., [126], [187]). Here we use the expression given in [278], with Dirichlet smoothing. The deviation from randomness scheme has recently been used with very good results at TREC, for the Web and Robust tracks ([7], [9]), and at CLEF, for Italian, Spanish and French monolingual tasks ([9], [8]). It is best described in [10].

The automatic extraction of index terms from full-text documents can also be seen as a feature selection problem, which has been investigated in data mining and machine learning (e.g., [147], [135]). The common approach is to choose the feature subset that maximizes the performance of a certain performance task or minimizes some form of error, but this might be too difficult in many cases. Even feature clustering can be an effective mechanism for addressing the problem of reducing the dimensionality within document corpora ([14], [72]).

Some forms of automatic indexing have been used in several information retrieval systems based on concept lattices (e.g., [37], [39], [44], [191], [54]). In [54], several simple indexing strategies for structured documents are used and a first attempt to measure their relative effectiveness is made. Automatic indexing has also been performed using a domain-specific thesaurus ([59], [60]) or WordNet® [127]. Clearly, reducing the set of features may affect the retrieval effectiveness, although this does not necessarily result in performance degradation. The effects of feature selection on concept lattice-based information retrieval are discussed in [39], for an interactive subject searching task, and in [44], for automatic text ranking.

Best-match text ranking has long been of interest in the information retrieval field ([245], [204], [13]) and it is still one of the most active research topics, with a steady growth in performance at TREC (e.g., [200], [7], [140]). The best-matching model is also used by current Web search engines to rank Web documents according to their relevance to the user query, together with link-based techniques.

To address the vocabulary problem, a number of approaches have been presented, including *relevance feedback* [119], general-purpose thesauri [249], *local feedback* ([270], [271]), collection-specific *lexical networks* [65], and *distributional retrieval feedback* ([43], [33], [47]).

Cluster-based retrieval has been justified on the grounds of the *cluster hypothesis* ([245], [123]), which states that relevant documents tend to be more similar to each other than non-relevant documents. Clustering algorithms, whether hierarchical or partitional, are described in several books and reviews (e.g., [76], [128]). The application of hierarchical clustering techniques to information retrieval is analysed in [265].

The use of interdocument similarity for improving text ranking is not specific to hierarchical clustering. In *latent semantic indexing* [71] and the *generalized vector space model* [268], for instance, the relationships between documents are exploited to transform the representation of the documents themselves before ranking. A notion of document space is also used, in a more abstract manner, in *topological ranking* [75].

Concept lattice-based ranking was proposed by Carpineto and Romano in [44]; the treatment offered here, although largely based on [44], is simpler and more elegant because it makes explicit use of the notions of query concept and document concept. In [44], a comparative evaluation of the retrieval performance of CLR, BMR, and HCR was also performed using the CACM and CISI test collections. The results showed that CLR was comparable to BMR and better than HCR on the whole document set, whereas CLR clearly outperformed the other two methods on ranking the documents that did not match the query.

As already mentioned, document clustering can be combined with best-match ranking. The application of clustering to retrieval results has been advocated in several papers (e.g., [123], [277]); this kind of clustering-based reranking or post-retrieval clustering has recently shown good retrieval performance ([141], [239]). Therefore, there seems to be much scope for the effective and efficient use of CLR on large document collections.

A concept lattice-based approach can also be applied to the results of conventional statistical clustering [127]. In this way, it is hoped to combine the efficiency of non-conceptual clustering with the clarity of conceptual clustering.

4

Text Mining

While text retrieval is concerned with finding the best elements among a variety of documents, following a query with the requested information on the part of the user, text mining involves the discovery of information previously unknown to the user. Concept lattices are suitable not only for text retrieval, as seen in the previous chapter, but also for text mining.

In this chapter we present two case studies. The first deals with mining the content of portions of the ACM Digital Library. We describe a concept lattice-based framework which allows hybrid search strategies (e.g., querying, browsing, bounding, thesaurus climbing) to discover information that would be difficult to acquire using conventional front ends.

The second case study concerns mining Web retrieval results using concept lattices. We present CREDO, a concept lattice-based system that allows visualization and navigation of the content of the documents returned by a Web search engine in response to a user query.

4.1 Mining the content of the ACM Digital Library

The first case study concerns the Digital Library of the Association for Computing Machinery (ACM). We will show that by using the concepts and techniques presented in the previous chapters it is possible to expand the functionalities currently provided by the ACM Digital Library to include support for interactive content mining and knowledge discovery.

The case study has the following structure. We first describe the main characteristics of the ACM Digital Library, including the use of the ACM Computing Classification Scheme (CCS) and the enhanced search and browse features recently introduced with the new ACM Portal. We briefly review the numerous advances of the new system and then point out some of its limitations, mainly the lack of user support for explorative data analysis and

Concept Data Analysis: Theory and Applications. Claudio Carpineto and Giovanni Romano
 ISBN: 0-470-85055-8

mining. In the rest of the section, we focus on this issue and address it by using concept data analysis.

We use as an illustration the set of papers published in Volume 18 (i.e., year 2000) of the ACM *Transactions on Information Systems* (TOIS). After showing the concept lattice built from the TOIS papers indexed with CCS terms, we present an interface which displays the resulting TOIS lattice to the user and allows him or her to query, browse, and bound the set of concepts in a simple and integrated manner. In this way, it is easy for the user to explore the actual content of the document collection, without needing to read single documents or even know the terminology of their subject domain. It is also possible for the user to discover trends and patterns difficult to detect with the existing search methods.

4.1.1 The ACM Digital Library

Works published by ACM since its inception have been digitized and indexed in a variety of ways into an on-line collection known as the ACM Digital Library. The library, hosted on the ACM website (http://www.acm.org), contains nearly half a century of fundamental research and serves a large and fast-growing user community, with about a million computer accesses to the Digital Library a month that result in the download of millions of full-text and metadata pages.

The works contained in the Digital Library have been grouped according to a number of categories (i.e., journals, magazines, transactions, proceedings, newsletters, publications by affiliated organizations and special-interest groups) and the user may browse the documents contained in each category to find those of interest. For instance, the user may access a certain ACM TOIS paper by selecting TOIS in the transactions category, then by choosing a volume and an issue within that volume, and finally by scanning the listing of articles in the selected issue. For each article, the Digital Library displays bibliographic information, abstracts, reviews, and full text.

In addition to browsing the Digital Library, the user may directly search the full text (or title, or abstract) of available publications by specifying which terms should be matched by the retrieved documents, with possible additional constraints to limit the scope of search to a subset of publications. For instance, the user may search for articles about John von Neumann's seminal contributions to computing which have appeared in journals by typing 'von Neumann' in the search box and restricting the search to the title fields of journal articles.

The browse and search modes just described have for some time been the only two ways to access the information published in the ACM Digital Library. More recently, in 2001, the ACM Portal (http://portal.acm.org) was released. The fundamental components of the ACM Portal are an enhanced version of the ACM Digital Library plus an extended bibliographic database consisting of more than a quarter-million citations of core works in computing. The introduction of the ACM Portal has also significantly reduced full-text download times and the response times for searching through a new system architecture and increased server capacity and bandwidth.

Most relevant to this case study, the ACM Portal has made a number of browsable *views* into the literature available. The established views are by author, by the ACM Computing Classification System (CSS), by subject, and by technical interest (i.e., with a user-selected and editable profile). Also, views exist side by side with search, including a quick search on all results pages and a find similar articles on a citation page.

The view by CSS is perhaps the most interesting and useful enhancement. The heart of the CSS (http://www.acm.org/class) is a subject tree. The tree consists of 11 first-level nodes and one or two levels under each of these. The first-level nodes have letter designations (A–K). The second and third levels have combination letter-and-number designations. For instance, the first-level node 'H. Information Systems' has five children, the first of which, 'H.1 Models and Principles', has, in turn, two children, 'H.1.1 Systems and Information Theory' and 'H.1.2 User/Machine Systems'.

The set of children of all first and second-level nodes begins with a node 'General' and ends with a node 'Miscellaneous'. In addition to the three-level tree enriched with the General and Miscellaneous nodes, the full classification scheme contains a fourth uncoded level, as well as general terms applicable to any elements of the tree that are relevant, and implicit subject descriptors.

The view by CSS takes advantage of the fact that all the works contained in the ACM Digital Library come with a manually provided set of CSS index terms, and that the bibliographic information contained in the ACM Portal's expanded database has been similarly indexed. A view by CSS allows the user to browse through the CSS's hierarchically structured subject tree from more general to specific categories. At all stages of such a drill-down, the user may access the listing of articles which are indexed by the category at hand, while context is explicitly maintained for the user; the underlying database query is constructed and displayed for enlargement or refinement.

All the tools discussed so far are very useful for finding which of the works contained in the Digital Library cover certain subjects; however, they are less effective at finding which subjects are covered in the works. This issue is dealt with in the next subsection.

4.1.2 Information retrieval and data view versus text mining

As both the ACM Digital Library and the expanded bibliographic database are fully indexed, the user may use the search mode to retrieve works which are potentially relevant to the information need expressed by a query. The views established by the ACM Portal play a similar role, with one important difference, namely that the documents are indexed and retrieved by using a controlled vocabulary.

This implies that the results are more reliable than those retrieved by automatic search, provided that the vocabulary concepts match the user need. Furthermore, the integration of search within view may facilitate view browsing and may also result in the retrieval of relevant documents that are not described by the same terms or concepts used in the view.

Using the currently available tools, a number of interesting search tasks can be easily accomplished. For instance, a view by author allows the user to drill down to a page that might be called an 'author's virtual bibliographic home page', listing all the works by that author known to the system. This listing might then be enhanced with additional works and projects provided by the author him- or herself. A view by CSS may be used to look into a specific domain, with the user drilling down from more general categories to narrower subjects and then to specific topics, thus progressively reducing the set of relevant results. For instance, when passing from 'operating systems' to 'process management' and then to 'threads', the set of results returned by the view shrinks from tens of thousands to a few thousand, and then to hundreds. Or, to take an example involving multiple search constraints, one can find how many books with 'software engineering' as their primary subject have been published by John Wiley & Sons in the last 10 years.

On the whole, the range of tools currently available for accessing the information contained in the Digital Library adequately support the user when he or she is interested in finding those documents which are described by certain terms or categories. By contrast, the same tools may be of little help if the user wants to discover the content of specific sections of the Digital Library or to mine the concepts contained in a set of articles which have been filtered out by using a particular criterion. This is an instance of the dichotomy between information retrieval and data view on the one hand and text data mining and concept exploration on the other.

More specifically, given a collection of documents *D*, information retrieval and data view are mostly ineffective at processing queries such as:

- Find the most common or uncommon subjects in *D*.
- Find which subjects imply, or are implied by, other subjects in *D*.
- Find novel and unpredictable subject associations in *D*.
- Find which subjects allow gradual refinement of subsets of *D*.

We will see in the following sections that these questions can easily be answered with a concept lattice-based method.

4.1.3 Constructing the TOIS concept lattice

The potential of the concept lattice-based methodology for disclosing the content of the ACM works will be discussed by using as an illustration a simple bibliographic database consisting of the 12 articles appearing in the four issues of Volume 18 of TOIS (year 2000). The entire processing chain will be illustrated; we will start from the raw data, which are enriched with thesaurus information, then go on to analyse the construction of the associated concept lattice, and finally describe a user interface to the previously constructed lattice that allows data mining and concept exploration.

The input data: Volume 18 of ACM TOIS + CCS

The input data consists of the set of articles published in Volume 18 of ACM TOIS. Each article is described by its bibliographic information and by its primary and additional classification terms, which belong to the ACM CCS; in this study we use only these index terms, although each TOIS paper may also contain a set of additional uncontrolled keywords, labelled just as 'keywords'.

Note also that ACM articles are usually provided with all index terms, including those terms that are implied by some other term according to the ACM CCS. In general, however, it is more convenient to assume that only some of the terms are manually assigned and that others are derived automatically.

The complete list of articles, along with their index terms, is shown in Figure 4.1.

In Table 4.1 we show the usual document–term relation associated with our example database, i.e., our input context. At this point, one could simply build the lattice associated with this context by using one of the algorithms presented in Section 2.1.3 or Section 2.2.1. However, in this case it is more convenient to try to also exploit the thesaurus information.

As already mentioned, the keywords used to index each document belong to the ACM CCS. Thus, each document can be seen as described not only by the keywords shown in Table 4.1 but also, in an implicit manner, by any of the ancestors of those keywords, according to the CCS. In other words, we may assume that the intended meaning of the partial ordering relation defined over the terms by the CCS is that any term implies any of its more general terms.

In Figure 4.2 we show the portion of the CCS that is relevant to our example; each term in the tree of Figure 4.2 is either a term used to index the documents in Table 4.1 or an ancestor of at least one of those terms. The tree can thus be used to expand the document representations. For instance, the article d1 is also implicitly described by the terms 'C. Computer Systems Organization', which is the father of 'C.2 Computer-Communication Networks' and 'C.4 Performance of Systems', and by 'H. Information Systems', which is the father of 'H.3 Information Storage and Retrieval'. Some of the terms derived from the CCS are completely new to the context, such as 'C. Computer Systems

d1: Evaluating the performance of distributed architectures for information retrieval using a variety of workloads, Cahoon, McKinley, and Lu, TOIS, 18(1), 1–43.
C.2 Computer-Communication Networks, **C.4** Performance of Systems,
H.3 Information Storage and Retrieval

d2: Shortest-substring retrieval and ranking, Clarke and Cormack, TOIS, 18(1), 44–78.
H.3.3 Information Search and Retrieval, **H.3.4** Systems and Software

d3: Improving the effectiveness of information retrieval with local context analysis, Xu, Croft, TOIS, 18(1), 79-112.
H.3.3 Information Search and Retrieval

d4: Fast and flexible word searching on compressed text, Silva de Moura, Navarro, Ziviani, and Baeza-Yates, TOIS, 18 (2), 113-139.
E.4 Coding and Information Theory, **H.3.3** Information Search and Retrieval

d5: Extending document management systems with user-specific active properties, Dourish, Edwards, LaMarca, Lamping, Petersen, Salisbury, Terry, and Thornton, TOIS, 18(2), 140–170.
C.2.4 Distributed Systems, **D.4.3** File Systems Management, **E.5** Files,
H.3.2 Information Storage, **H.3.3** Information Search and Retrieval,
H.3.4 Systems and Software

d6: Efficient content-based indexing of large image databases, El-Kwae and Kabuka, TOIS, 18(2), 171–210.
C.2 Computer-Communication Networks, **C.4** Performance of Systems,
H.3 Information Storage and Retrieval

d7: Chimera: hypermedia for heterogeneous software development environments, Anderson, Taylor, and Whitehead,, TOIS, 18 (3), 211–245.
D.2 Software Engineering, **H.5** Information Interfaces and Presentation,
I.7.2 Document Preparation

d8: The maximum entropy approach and probabilistic IR models, Greiff and Ponte, TOIS, 18(3), 246–287.
H.3.3 Information Search and Retrieval

d9: Data integration using similarity joins and a word-based information representation language, Cohen, TOIS, 18(3), 288–321.
H.2.3 Languages, **H.3.3** Information Search and Retrieval

d10: Model-driven development of Web applications: the AutoWeb system, Fraternali, Paolini, TOIS, 18(4), 323–382.
H.5 Information Interfaces and Presentation, **D.2** Software engineering

d11: Beneath the surface of organizational processes: a social representation framework for business process redesign, Katzenstein and Lerch, TOIS, 18(4), 383–422.
K.4 Computers and Society

d12: Beyond intratransaction association analysis: mining multidimensional intertransaction association rules, Lu, Feng, and Han, TOIS, 18(4), 423–454.
H.2.8 Database Applications

Figure 4.1 Example database consisting of the articles published in ACM TOIS, Volume 18, 2000, along with their index terms.

Table 4.1 The TOIS 2000 context

	d1	d2	d3	d4	d5	d6	d7	d8	d9	d10	d11	d12
C.2	×				×							
C.2.4					×							
C.4	×											
D.2							×			×		
D.4.3					×							
E.4				×								
E.5					×							
H.2.3									×			
H.2.8						×						×
H.3	×	×	×	×	×	×		×	×			
H.3.1						×						
H.3.2					×							
H.3.3		×	×	×	×	×		×	×			
H.3.4		×			×							
H.5							×			×		
I.7.2							×					
K.4											×	

Organization'; others may be used to describe different documents in the original context; e.g. 'H. Information Systems' in d12.

Building the concept lattice from the raw data

We saw in Section 2.4.1 that it is possible to take advantage of the presence of a thesaurus defined over the index terms by incorporating this information into the concept lattice in such a way that more general terms index more general classes.

The concept lattice built from the relation in Table 4.1 enriched with the thesaurus information shown in Figure 4.2 is displayed in Figure 4.3. To construct the lattice, we must adapt one of the lattice-building algorithms (e.g., the Update by Local Structure algorithm) to take into account the thesaurus information, as explained in Section 2.4.1.

The concepts in the lattice in Figure 4.3 are labelled using minimal labelling. As explained in Section 1.2, the intent of each concept inherits the term labels from its most general concepts and the extent inherits the document labels

C. Computer Systems Organization
    C.2 Computer-Communication Networks
        C.2.4 Distributed Systems
    C.4 Performance of Systems

D. Software
    D.2 Software Engineering
    D.4 Operating Systems
        D.4.3 File Systems Management

E. Data
    E.4 Coding and Information Theory
    E.5 Files

H. Information Systems
    H.2 Database Management
        H.2.3 Languages
        H.2.8 Database Applications
    H.3 Information Storage and Retrieval
        H.3.1 Content Analysis and Indexing
        H.3.2 Information Storage
        H.3.3 Information Search and Retrieval
        H.3.4 Systems and Software
    H.3 Information Interfaces and Presentation

I. Computing Methodologies
    I.7 Text Processing
        I.7.2 Document Preparation

K. Computing Milieux
    K.4 Computers and Society

Figure 4.2 Portion of the ACM CCS relevant to the articles in Figure 4.1.

from its most specific concepts. For instance, the unlabelled concept furthest to the left in Figure 4.3 has the intent {H, H.2, H.3, H.3.3} and the extent {d9, d6}. It should be noted that the minimal labelling representation is used here for the sake of concise visualization; the actual user interface will make use of a much more explicit concept representation (see below).

The lattice in Figure 4.3 is deep and rather narrow. The document concepts of seven out of the 12 documents are atoms (i.e., they are upper neighbours of the bottom element of the lattice), as one might expect; however, the document concepts of the remaining five documents are superconcepts of concepts other than the bottom element, reflecting the fact that some documents have been indexed using a supercategory of keywords used to index a different document. The relatively high degree of structuring in the lattice is due to the presence of many CCS supercategories which are shared by several articles, either explicitly or implicitly.

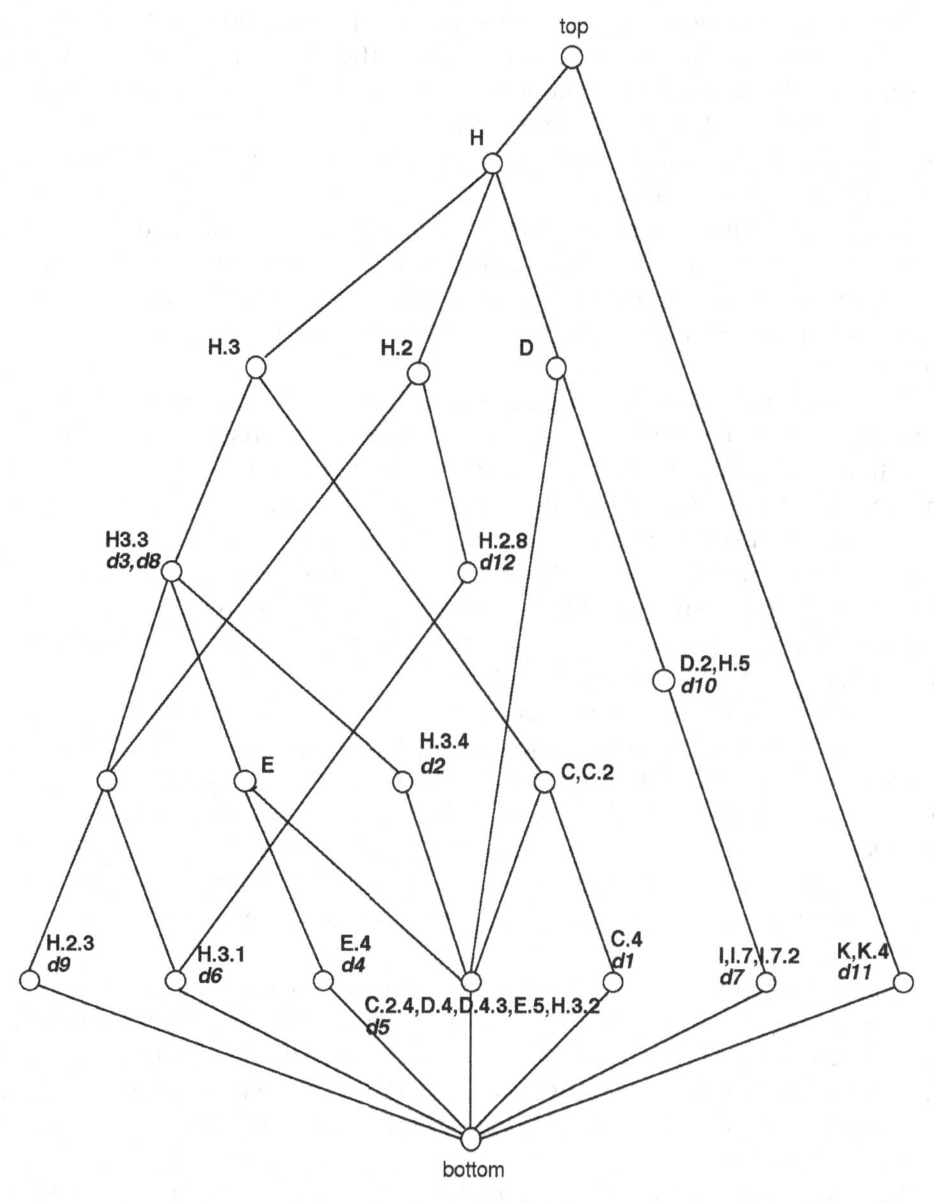

Figure 4.3 The concept lattice built from the context in Table 4.1 and the thesaurus in Figure 4.2.

Discovering relationships between subjects

The TOIS lattice allows fine-grained exploration of the collection content. Among the kind of questions that can easily be answered by inspecting this particular representation, as opposed to conventional information retrieval and data view approaches, are the following.

Common or uncommon subjects. There are no terms shared by all documents (i.e., the intent of the top element of the lattice is empty). The term 'H. Information Systems' is common to a vast majority of the documents, as expected; in fact, all the documents but d11 are described by it. The terms describing d11 ('K. Computing Milieux' and 'K.4 Computers and Society') do not match any other document.

Subjects which imply, or are implied by, other subjects. As explained above, the lattice structure reflects the semantic implications contained in the CCS. This means that each concept may be generalized (specialized) by replacing one of the terms contained in its intent with a more general (specific) term according to the CCS.

In addition, the lattice may reveal other implications that are specific to the context at hand. For instance, the TOIS lattice shows that it is not only the case that 'K.4 Computers and Society' implies (→) 'K. Computing Milieux' (as imposed by the CCS), but also that K → K.4 (because the most general node that contains K contains also K.4). The same goes for I → I.7 → I.7.2 and D.4 → D.4.3. The fact that two terms which are ordered by a more general than relation (according to the CCS) turn out to be interchangeable in the concept lattice may reveal that their granularity is too fine to describe the content of the particular collection at hand.

The lattice might also show that an implication holds between two terms that are unrelated according to the CCS. For instance, the following implications hold in the TOIS lattice, among others: 'D.2 Software Engineering' → 'H. Information Systems', and 'C. Computer Systems Organization' → 'H.2.8 Database Applications' (as will be better seen in Chapter 5, an implication $t_1 \rightarrow t_2$ holds whenever the most general concept containing t_1 is more specific than (i.e., is a subconcept of) or equal to the most general concept containing t_2).

This may suggest that the two terms involved in the implication are related to each other according to the characteristics of the document collection being explored, because the implication antecedent may be seen as a typical application area of the technique expressed by the consequent for instance, or because the antecedent is an emerging topic of an old research area described by the consequent.

Novel and unpredictable subject associations. The discovery of novel research subjects or application areas in a certain scientific context may be of interest to many users. This kind of request cannot easily be translated into a query matching the given indexing language; in fact, it seems to require some form of direct inspection of the database. The use of concept lattice plus bounding (described in Section 2.4.2) can effectively support this search task.

For the TOIS case, for instance, after realizing that most documents are about 'H. Information Systems', the user might prune all the concepts with

'H. Information Systems' from the concept lattice, thus restricting the search space to only one node, namely the concept with intent {'K. Computing Milieux, K.4 Computers and Society'} and extent {d11}. This does indeed seem a novel subject for TOIS; in fact, paper d11 is not truly about traditional information systems, in that it describes a representation framework for organizational process design.

For the discovery task, browsing through the lattice nodes can also be very useful. For instance, the path linking 'H. Information Systems' to 'D. Software' to 'D.2 Software Engineering' down to 'I. Computing Methodologies, I.7 Text Processing, I.7.2 Document Preparation' may suggest that *software* and, to a smaller extent, *software engineering* can be considered as classical application areas of *information systems*, and that *document preparation* may be also considered as a specific topic of those broad application areas, at least in the TOIS setting. In fact, paper d7 describes a data modelling approach for supporting software development.

Subjects for query refinement (or enlargement). As seen in the previous chapter, following edges departing upwards (downwards) from a query concept produces all minimal conjunctive refinements (enlargements) of the query with respect to that particular database, whether unexpanded or expanded with a thesaurus. In practice, the lattice suggests which subjects should be added to (or deleted from) those describing a subset of documents to yield such minimal refinements (enlargements).

For instance, the node with intent {H. Information Systems}, which contains 11 documents, can be minimally refined in three possible ways: {H. Information Systems, H.3 Information Storage and Retrieval}, {H. Information Systems, H.2 Database Management}, and {H. Information Systems, D. Software}, which contain 8, 6, and 3 documents, respectively.

It should be noted that D, which is a subject producing a minimal refinement of H, is not related to H according to CCS. Any other possible refinement will either produce a more specific concept than at least one of the three minimal refinements, or would be inadmissible with respect to the given collection. Each of the first two minimal refinements corresponds to a slightly narrower topic of the information systems subject, whereas the third refinement can be seen as specifying a broad application domain for the information systems subject.

This example and those given above show that a concept lattice can be used to analyse a general semantic hierarchy like CCS 'in context', with the goal of identifying inconsistencies and anomalies that arise from its concrete applications.

In the next subsection we will address the issue of the retrieval interface to the concept lattice, i.e., the practical means for the user to take advantage of the information disclosed by its nodes and structure.

4.1.4 Interacting with the TOIS concept lattice

The second step of the methodology illustrated by the ACM Digital Library case study is a retrieval interface to the concept lattice built in the earlier step. The user interface consists of four main components, namely one visualization module and three interaction modules, for browsing, querying and bounding.

From an implementation point of view, the concept lattice associated with a specific document collection could be computed and saved to disk by a separate application, whereas the interface program would be invoked with appropriate parameters to specify the concept lattice to which it should be applied.

Such an interface program could be implemented as a *Common Gateway Interface* (CGI), running on a Web server. A CGI permits interactivity between a client and a host operating system through the World Wide Web via the *Hypertext Transfer Protocol* (HTTP). It is a standard for external gateway programs to interface with Web servers. In our case, the CGI interface program would manage the interaction between the user and the concept lattice by accepting user commands issued from a Web browser and computing the response to such commands using the concept lattice stored on the server.

In this subsection we describe each facility of the user interface in detail, and show how they can be combined to form hybrid text mining strategies.

Visualization

The first problem in the interface design is the visualization of the retrieval space. In general, the lattice representation of a document collection is too large to fit on a screen, even for small parts of the ACM Digital Library. As seen in Section 2.3, several techniques can be used to visualize concept lattices, depending on the application characteristics.

Here we have taken an approach based on focus + context views, assuming that there is a current focus (or region) of interest in the underlying concept lattice. This choice was in part motivated by the fact that in order to perform text mining the user will typically want to focus on some selected node and browse through the nodes in the adjacent region, which are likely to contain similar information, and in part by the availability of several natural mechanisms for selecting the current focus on the part of the user (e.g., querying, bounding, thesaurus climbing), as explained below.

The selection of a focus of interest causes a small region around the focus to be displayed that contains just its children. In keeping with our example, Figure 4.4 shows the initial screen displaying the most general concepts with a non-empty intent in the TOIS lattice (i.e., this can be seen as the default focus prior to the interaction of the user), along with their children. The intent of each focus node is displayed in full, along with the number of documents it contains, whereas for its children we do not show the complete list of terms

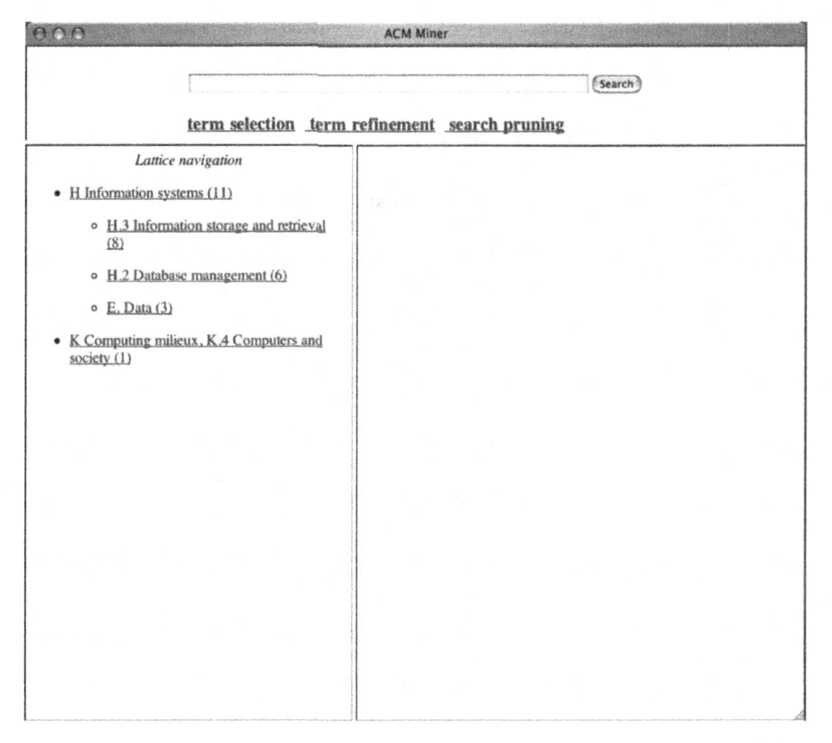

Figure 4.4 Initial screen displaying the most general concepts in the TOIS lattice of Figure 4.3.

describing each node. Rather, we use an indented representation to suggest how their intents can be derived starting from the focus node.

More precisely, for each child we display which term or terms should be added to the focus intent to obtain that node, as in Figure 2.22. In this way, we highlight which changes to a given node cause a transition to one of its children, which might be useful for driving query refinement. This choice also has the advantage that the no redundant term information is shown on the screen. Similar to the focus node, we display for each child the number of documents it contains. It should be noted that in our case the concept with intent {K, K.4} has no children.

By clicking on any displayed concept, the user may visualize in the lower right-hand frame of the interface screen the full intent of the concept and the list of documents it contains, in the form of bibliographic information. For each document, its abstract and full text can then be displayed on request (see, for instance, Figures 4.5 and 4.6).

We should emphasize that as the interaction between the user and the system can be conveniently handled by a Web-based client-server system, the actual screens returned by the user interface will be rendered as HTML pages. In fact, Figures 4.4–4.6 have been produced using HTML.

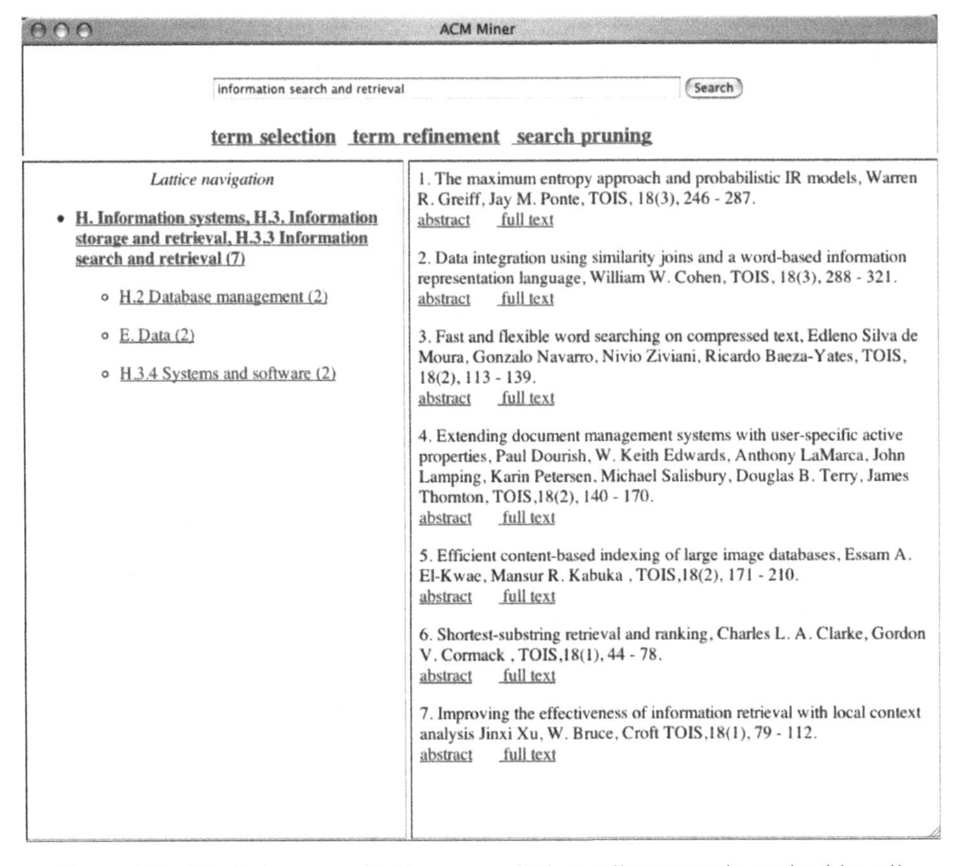

Figure 4.5 Display screen for the query 'information search and retrieval'.

Browsing

Once a focus has been selected, the user may navigate through the descendants of the displayed nodes by opening their downward links. As long as the focus of interest remains the same, the new nodes are added to the old ones, thus maintaining full contextual information (see, for example, Figure 4.6). This may be useful, for instance, when the user wants to navigate the results of a query. A new query will yield a new focus of interest, and the process may be iterated.

The properties of concept lattices support effective recognition of useful documents and/or subject terms while browsing. The intensional description of concepts and the minimality of query refinement are two such properties. Another useful feature is that the lattice structure, in which there are many paths to a particular node, facilitates recovery from bad decision making while traversing the hierarchy in search of documents, as opposed to the strict hierarchical structures used in most cluster-based browsers, in which each class has exactly one parent.

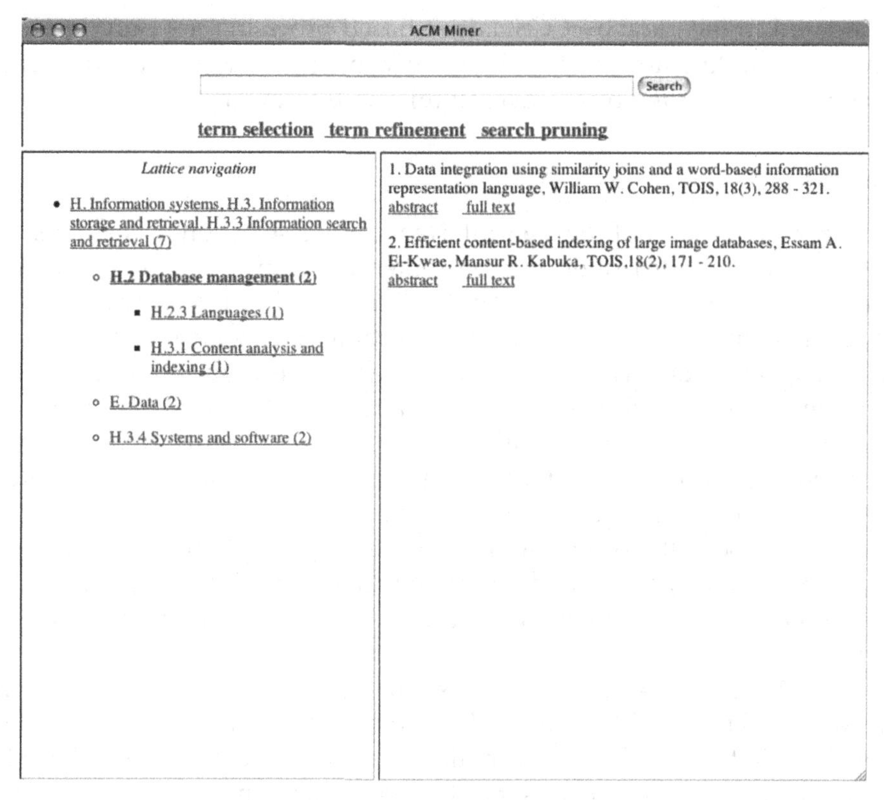

Figure 4.6 Display screen after the selection of the first child in Figure 4.5.

Furthermore, as the same document is often relevant to two or more queries that happen to have incomparable descriptions, the ability to deal with non-disjoint clusters is an important feature of browsing retrieval systems; lattice conceptual clustering naturally supports this functionality, as opposed to hierarchical conceptual clustering.

It is interesting to compare the browsing of the concept lattice to the view by CSS currently provided by the ACM Portal. While the latter does go some way towards achieving controlled query refinement, it does not allow the user to refine an entire concept describing a set of documents but only single keywords shared by those documents. This may result in a coarse refinement of a given set of documents, whereas lattice browsing guarantees minimal refinement of each subset of documents in the collection by allowing for simultaneous refinements of its describing keywords.

For instance, it should be noted that if we replaced H by H.5 we would not get a maximal refinement because the most general node containing H and H.5 is the node with intent {H, H.5, D, D.2}, which is indeed a subconcept of the minimal refinement {H, D}. Similarly, in order to refine the most general node with intent {H.2} one should use H.3.3 prior to H.2.3.

Compared to the view by CCS, lattice-based browsing not only improves granularity of refinement, but is also tightly integrated with the other search functionalities provided by the overall mining framework.

Querying

The background menu of the user interface allows selection of the other two basic interaction modes: querying, presented here, and bounding, discussed in the next section. A new query can be formulated in two ways: either the user specifies the new terms by using the command 'term selection', or the user modifies the current query (i.e., the intent of the current node) by using the command 'term refinement'. In the latter case the user may remove a term, add a new term, or specialize/generalize an existing term using the information contained in the CCS thesaurus. In either case, the new query will automatically appear in the search box.

The query mode allows the user to make large jumps to regions of interest. The result of a query is the node of the lattice (whose intent is) equal to the query, if there is any, or the most general of the nodes that are more specific than the query.

To illustrate, Figure 4.5 shows the system's response to the query 'information search and retrieval'. The node 'H. Information systems, H.3 Information Storage and Retrieval, H.3.3 Information Search and Retrieval' is selected as current focus (shown in bold), as it is the most general node among the nodes that are more specific than the user query.

The system also shows the documents associated with the focus node and its three children, along with the cardinality of the extent of each displayed node. It should be noted that the sum of the documents contained in its children (6) does not coincide with the number of documents in the focus node (7), in that two documents belong only to the focus node (i.e., d3 and d8) and of the remaining five documents, one (d5) is shared by two children. If the user points and clicks with the mouse on the first child, the system produces the screen shown in Figure 4.6. As remarked upon above, in this case the general focus has remained unchanged while the current node, including the list of displayed documents, is the one that was most recently selected by the user.

Bounding

The 'search pruning' command allows users to change the space from which they are retrieving information during the search. As illustrated in Section 2.4.2, the user may apply constraints (expressed as order-theoretic operators over a particular conjunction of terms) with which the sought documents have to comply and the retrieval space is bounded accordingly.

As a result, the partitions induced over the search space by the application of such constraints present useful properties from the point of view of the retrieval performance, as measured by *recall* and *precision*. Recall is defined as the ratio of number of items retrieved and relevant to the number of items relevant; precision is the ratio of number of items retrieved and relevant to the number of items retrieved. Recall measures the ability to retrieve all relevant documents, while precision measures the ability to retrieve only relevant documents.

Let c^* be the target concept and c_1 be a concept of the lattice. The interpretation of the four types of constraints is as follows.

$\uparrow c_1$ (*enlargement*): Concept c^* must contain at least the documents contained in c_1 (dually, c^* must not contain terms that are not contained in c_1). In terms of recall/precision, this implies that the precision of c_1 is equal to 1 and that we fix a threshold (i.e., the recall of c_1) below which the target recall cannot drop.

$\downarrow c_1$ (*refinement*): Concept c^* must contain at least the terms contained in c_1 (dually, c^* must not contain documents that are not contained in c_1). In terms of recall/precision, this implies that the recall of c_1 is equal to 1 and that we fix a threshold (i.e., the precision of c_1) below which the target precision cannot drop.

$\neg \uparrow c_1$ (*discrimination*): Concept c^* must contain at most a strict subset of the documents contained in c_1 (dually, c^* must not contain a strict subset of the terms contained in c_1). In other words, there is at least one non-relevant document in c_1 (i.e., the precision of c_1 is less than 1).

$\neg \downarrow c_1$ (*differentiation*): Concept c^* must contain at most a strict subset of the terms contained in c_1 (dually, c^* must not contain a strict subset of the documents contained in c_1). In other words, there is at least one term in c_1 that needs be suppressed (i.e., the recall of c_1 is less than 1).

Bounding the search space has of course a direct effect on browsing and querying, in that it allows the user to jump to only those nodes that are within the admissible region, but it may also change the space visualization. This happens whenever the current focus, as an effect of the new constraint(s), is no longer admissible; in this case the system makes the node(s) of the nearest boundary set the new focus.

Querying and bounding have different scopes. In both interaction modes the user provides some information about the goal, and the system focuses the search on some relevant space region; but there are two main differences. The first is that the description languages are usually different, and therefore some information may be expressed in one mode but not in the other. For instance, by applying the constraint $\neg \downarrow c_1$ we can specify the target nodes using negated terms, which is forbidden in the strictly conjunctive language of the lattice-based query mode.

The second difference is that the two strategies take advantage in different ways of the same piece of information. Suppose that users are interested in the documents containing c_1. Users may query the system by questioning c_1, or they may apply the constraint $\downarrow c_1$. In both cases, the likely result is a jump to a node containing c_1, but in the bounding mode the change will also affect later retrieval.

The advantage of bounding is that the whole search becomes more focused; the disadvantage is that the user is no longer able to retrieve relevant documents in the region that has been pruned away. Therefore bounding is more useful when the goal is very precise or when the user becomes more aware of the terminology/structure of the database, later on during the search.

An additional advantage of bounding is that the documents violating the constraints are removed from all the nodes in which they are contained, possibly including admissible nodes, thus reducing the time needed to scan the documents associated with the remaining nodes.

Putting the pieces together: forming hybrid mining strategies

At this point, it should be clear that at any given time the system is in a certain state, characterized by a current retrieval space and by a focus with a current node within it. In each state, the user may select an operator (browsing, querying, thesaurus climbing, or bounding) and apply it. As a result, a transition is made to a new state, possibly characterized by a new retrieval space and/or new focus and/or new current node. The new state is evaluated by the user for document retrieval, and then the whole cycle may be iterated.

Therefore each interaction sequence may be composed of several operators, connected in various ways. For instance, the user may initially bound the search space exploiting their knowledge about the goal, then query the system to locate a region of interest within the bounded space, then browse through the region; also, at any time during this process, the user may take advantage of the feedback obtained during the interaction to make a jump to a different but related region (e.g., by thesaurus climbing), or to further bound the retrieval space.

One distinguishing feature of our approach to interactive text mining is that the user does not have to map different representations and results while using different search methods, and there is no need for the system to maintain different structures. Well-known techniques such as querying, browsing and thesaurus climbing are integrated with novel techniques such as bounding into a unique search space including both terms and documents.

This is a deep and effective form of integration, because the search strategies share the same data space and exchange their search results. The advantage is that the user has at his or her disposal a range of functionalities which cater for a number of different tasks and constraints and that their effects may be combined to make the mining faster, more accurate, or more useful.

Before concluding this case study, we should emphasize that, although for the sake of illustration we chose a very simple database, our methodology can be applied with minor changes to small to medium-size text collections. Thus, content mining based on concept lattices can be applied not only to single journal volumes but also to multiple volumes or entire journals.

Equally important, it can also be applied to large text collections to post-process a set of results which have been extracted using a fast but inaccurate method. For instance, one might be interested in exploring the content of the papers authored by a certain person, or the papers published at a given time by several journals. Even the content of the documents returned in the results page in response to a user-selected view may be made accessible by the same technique. Thus, there seems to be a lot of scope for application of conceptual data analysis to interactively mining the content of the ACM Digital Library as well as other important document collections.

4.2 Mining Web retrieval results with CREDO

The second case study is concerned with mining the retrieval results returned by a Web search engine using a concept lattice approach. The case study has the following structure. We will first discuss the lack of effectiveness of current search engine interfaces and some solutions for overcoming this problem. Then we describe the design and implementation of CREDO, a system that allows the user to query Web documents and see retrieval results organized in a browsable concept lattice. Finally, we illustrate the operation and the utility of CREDO with some example sessions. An on-line version of CREDO is available for testing at http://credo.fub.it/.

4.2.1 Visualizing Web retrieval results

Current Web search engine interfaces are limited by a lack of a concise representation of the content of all retrieved documents. Conventional textual displays take much perusal time and screen space and do not enable inspection of multiple documents at the same time. Compounded with the unmanageably large response sets of Web search engines, with their low precision, the perusal of document summaries may be ineffective, time-consuming and costly for the user.

To alleviate this problem, new mechanisms that provide visual cues for interpreting retrieval results and extend user control over their presentation and selection are necessary. A well-known approach, among several others briefly discussed in Section 4.3, is to use hierarchical clustering, introduced in Section 3.2.1. The clusters built from Web documents are labelled using the document descriptions and the resulting structure is offered to the user for browsing, starting from the top clusters.

The effect for the user is a combination of query-based and category-based Web search methods. Query-based Web search (e.g., Google) is fast and usually effective, but requires that the user's information needs are expressed in an appropriate manner. Category-based Web search (e.g., Yahoo!) is also suitable for retrieving items that are unknown to the user or items that a particular user is not able to describe through a list of terms; however, the categorizing is usually produced manually and the user must adopt the mindset of those who created it.

Clustering retrieval results seems an effective way to combine these two approaches without retaining their main disadvantages. It allows the user to focus on some weakly specified subject (by a query), and then drill down through the hierarchy that has been created on the fly in response to that particular query.

However, hierarchical clustering of Web results also presents some drawbacks. First, as already discussed in Section 3.2.1, valuable clusters can be omitted in the resulting hierarchy due to the use of similarity metrics and heuristic choices during cluster formation. Second, once the hierarchy has been constructed, a label for each cluster must be found. This step is very important for the task at hand, because the labels guide the process of cluster selection and subsequent refinement on the part of the user. However, it is often difficult to find a good description for a set of documents. Finally, the user navigates through a strict hierarchy, which does not easily permit recovery from bad decisions.

Most of these limitations can be overcome using concept lattices. The set of clusters is complete and each element is formally justified, cluster labelling is integrated with cluster formation, and the structure is a lattice instead of a hierarchy. In addition to being theoretically appealing, on-line mining of Web results using concept lattices is technically feasible, given the state of the art of concept data analysis and the advances in Web-based programming. These considerations have led to the development of the CREDO system, described in the next subsection.

4.2.2 Design and implementation of CREDO

CREDO, which stands for Conceptual REorganization of DOcuments, takes a user query as input. The query is forwarded to an external Web search engine, and the first pages retrieved by the search engine in response to the query are collected and parsed. At this point, a set of index terms that describe each returned document is generated. Such terms are next used to build the concept lattice of the retrieval results. The last steps consist of visualizing the lattice and managing the subsequent interaction with the user, who may browse through the concepts and display the associated documents.

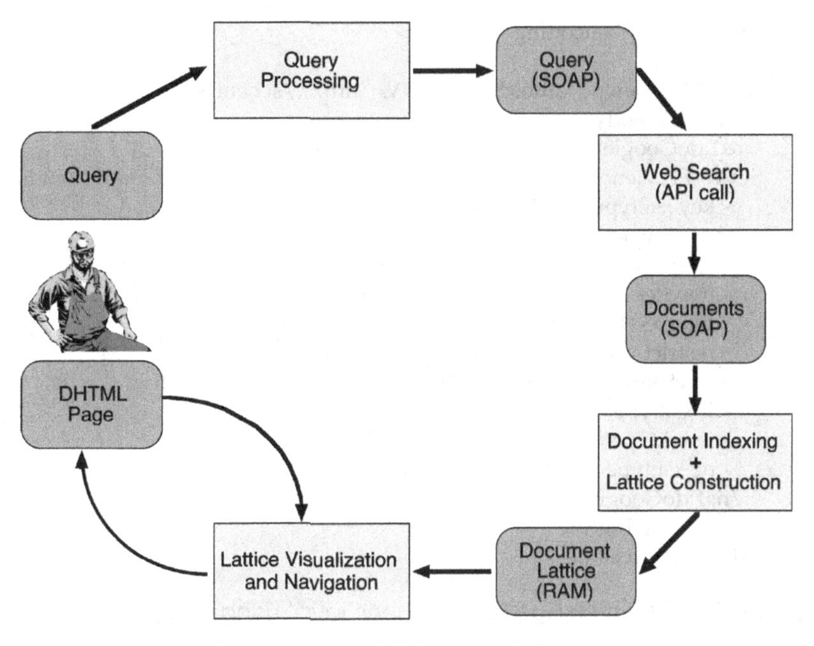

Figure 4.7 CREDO's architecture.

CREDO is a client-server system implemented in *Shark*, a Lisp-like language being developed by the second author of this book that is automatically translated into ANSI C and then compiled using GCC. CREDO runs on a Web server as a CGI application that can be invoked from any Web browser. In its current version all the main components run on the server, although more balanced choices are conceivable, depending on the requirements of the visualization and interaction steps.

Figure 4.7 shows the architecture of CREDO, as summarized above (functions are described in squared boxes, outputs in rounded boxes). In the following sections we detail the working of its main blocks.

Interaction with the search engine

We assume that the interaction between CREDO and the search engine is handled through SOAP messages (http://www.w3.org/TR/SOAP). SOAP is a short for Simple Object Access Protocol, a lightweight messaging protocol used to encode the information in Web service request and response messages before sending them over a network. SOAP messages are based on Extensible Markup Language, or XML (http://www.w3.org/TR/REC-xml), a human-readable, machine-understandable, general syntax for describing data. As SOAP messages are independent of any operating system and may be transported using a variety of Internet protocols, including HTTP, they are widely used for exchanging structured and typed information over the Web.

```
<?xml version='1.0' encoding='UTF-8'?>

<SOAP-ENV:Envelope xmlns:SOAP-ENV="http://schemas.>
  <SOAP-ENV:Body>
    <ns1:doGoogleSearch xmlns:ns1="urn:GoogleSearch"
    SOAP-ENV:encodingStyle="http://schemas.xmlsoap.org/soap/encoding/">
      <key xsi:type="xsd:string"><key>
      <q xsi:type="xsd:string">java</q>
      <start xsi:type="xsd:int">0</start>
      <maxResults xsi:type="xsd:int">10</maxResults>
      <filter xsi:type="xsd:boolean">true</filter>
      <restrict xsi:type="xsd:string"></restrict>
      <safeSearch xsi:type="xsd:boolean">false</safeSearch>
      <lr xsi:type="xsd:string"></lr>
      <ie xsi:type="xsd:string">latin1</ie>
      <oe xsi:type="xsd:string">latin1</oe>
    </ns1:doGoogleSearch>
  </SOAP-ENV:Body>
</SOAP-ENV:Envelope>
```

Figure 4.8 SOAP message for the query 'java' using Google Web APIs.

In our case, we need to encode the information concerning the query and the retrieval results. An example of search engine that accepts queries from an authorized computer program using this protocol is Google, through its Web APIs service (http://www.google.com/apis/).

To illustrate, suppose that we want to question CREDO with the query 'java'. After the query has been typed in, a SOAP call to the search engine is generated. In Figure 4.8, we show the SOAP call for the query 'java' using Google Web APIs. Note that in addition to the query terms, contained in the 'q' element, other useful pieces of information are specified, including the number of result pages required (in the 'maxResults' element).

CREDO collects the first 100 retrieved documents. This number is a good compromise between having few documents (which can be better examined sequentially and are possibly too homogeneous in content to give rise to useful subgroups) and a large number of documents (which are possibly not relevant and cannot be processed with reasonable response times).

The output of the search engine is returned as a SOAP message, analogous to that received as an input, which CREDO transforms into a tree-based internal representation. The result of this operation for the query 'java', limited to the first retrieved document, is shown in Figure 4.9. Note that several attributes of interest for the extraction of a subset of index terms are present, including a 'title' and 'snippet' for each retrieved document.

It is also possible to query a search engine in a direct manner via the HTTP protocol, and then acquire the results by parsing the HTML pages returned. This approach has the advantage that the utilization of the search engine may

```
(XML
  (HEADER "<?xml version='1.0' encoding='UTF-8'?>")
  (BODY (SOAP-ENV_ENVELOPE
    (ATTLIST ...)
    (SOAP-ENV_BODY
      (NS1_DOGOOGLESEARCHRESPONSE
        (ATTLIST (ATTLIST (NAME "ns1:doGoogleSearchResponse") ...))
        (RETURN (NS1_GOOGLESEARCHRESULT
          (DOCUMENTFILTERING "true")
          (ESTIMATEDTOTALRESULTSCOUNT "37,800,000")
          (DIRECTORYCATEGORIES..)
          (SEARCHTIME "0.216056")
          (RESULTELEMENTS
            ((ITEM
              (ATTLIST ...)
              (CACHEDSIZE "30k")
              (SNIPPET "featuring developer resources ...")
              (RELATEDINFORMATIONPRESENT "true")
              (DIRECTORYTITLE .)
              (URL "http://java.sun.com/")
              (TITLE "Java.Sun.com"))
            ...
```

Figure 4.9 Tree-based internal representation of the SOAP message returned by Google in response to the query 'java' (limited to the first retrieved document).

not be subject to external constraints. However, the parsing process is prone to error and is sensitive to changes in the search engine interface.

Indexing of retrieval results

One possibility is to use the method described in Section 3.1.3, considering as the collection at hand either the whole Web or just the first retrieved documents. As in our case the documents of interest have been retrieved from a larger collection in response to a user query, in order to discriminate between good and poor terms it may be more convenient to compare occurrence in the retrieved documents with occurrence in all documents.

In particular, it may be expected that the frequency of good terms will be higher in the retrieved documents (implicitly considered relevant) than in the whole collection, while other terms will occur with the same frequency in both document sets. This approach is usually referred to as *distributional term weighting*, because we use term-scoring functions based on the difference between the distribution of the terms in the set of retrieved documents and the distribution of the terms in the entire collection.

The best known of such functions is the Kullback–Leibler distance:

$$\text{score}(t) = p_d(t) \cdot \log \frac{p_d(t)}{p_C(t)}$$

where t is a term, d is a document, $p_d(t)$ is the probability of occurrence of term t in document d, and $p_C(t)$ is the probability of occurrence of term t in the whole collection. The probabilities can be estimated using the frequency of terms, and a large reference document collection can be used instead of the whole Web to compute $p_C(t)$.

Using the Kullback–Leibler distance, the scores assigned to each term may more closely reflect the relevance of the term to the specific query at hand rather than the general importance of the term in the collection.

Both the conventional and the distributional term weighting methods can be implemented efficiently and usually exhibit good retrieval performance, but they require that the description of each document should be detailed enough to have a rich representation for the document's language model.

Unfortunately, this is not the case in our application. The document summaries are too short, whereas the full-text documents are not available unless we download the original documents from the Web (which may take too long) or have access to cached copies of the documents (like some search engines). On the other hand, we can take advantage of the document structure.

The solution adopted by CREDO takes into account these practical constraints, considering just the information contained in the results returned by the search engines and focusing on the elements that best describe the content of the documents. Each document is indexed by two sets of terms, one for the title and one for the snippet, extracted from the values of the corresponding attributes in Figure 4.9. The cleaning procedure is the same for both attributes; it essentially consists of (i) identifying all tokens formed by alphanumerical characters, (ii) conversion of upper-case characters to lower-case, (iii) stripping numbers, and (iv) removing stop words.

Lattice construction and visualization

One of the main difficulties is that the concept lattice of the retrieved documents may contain many irrelevant concepts resulting from spurious combinations of the document terms. This problem is especially relevant to the set of coatoms (i.e., the lower neighbours of the top element of the lattice), which must be shown to the user at the beginning of the interaction and should give an immediate idea of the main subjects into which the results can be grouped.

To address this problem, one can use a very limited number of terms per document (e.g., the title of the document). This has the advantage that we reduce the possibility that two documents that are different in content share some word by chance. However, by doing so it is likely that many documents will not share any term with the other documents, thus remaining ungrouped (i.e., their document concept will be directly linked to the top and the bottom elements of the concept lattice).

To avoid this drawback, one can take a hybrid approach in which the lower levels of the lattice are built using a larger set of terms than those used to build the top level.

CREDO implements this strategy in the following manner. To build the top level of the lattice, consisting of the top element of the lattice along with the coatoms, it uses only the title terms (if some query term is not contained in the title, which is not usually the case for Web searches, it is added to the title to make sure that there is a natural starting point for browsing the document lattice). Then the coverage of each coatom is increased by including the documents that contain the concept terms in their snippet. This amounts to expanding the description of each document with those snippet terms which are also concept terms. At this point, the lower levels are built using the expanded document representation.

This can be seen as a two-step classification procedure, in which the first layer identifies the main topics and the other layers contain the subtopics of each main topic. Thus, for example, in response to the query 'Washington' we may have a top classifier, which distinguishes documents about *Washington DC* from documents about *Washington State* or *Washington University*, and lower classifiers, which distinguish *lodging* from *transportation* in *Washington DC* (or in *Washington State*, or *Washington University*).

Note that in this example *lodging* and *transportation* (or even *map*, or *FAQs*) are shared by more than one main topic. Using the document title only reduces the chance of obtaining such secondary topics at the top level as a cross-topic concept of little utility.

Clearly, the resulting clustering structure is not a true concept lattice, in the sense that it cannot be seen as the concept lattice of a specific context. In particular, it is not equivalent to the concept lattice which would be obtained if we indexed each document by taking the union of the terms contained in the title and in the snippet.

The implementation of the hybrid strategy described above is straightforward. CREDO makes use of the Next Neighbours algorithm described in Section 2.1.3, with some modification to account for the expansion of the document representation. More precisely, after finding the lower neighbours of the top element of the lattice using only the title terms as document intents, the context is expanded as specified above, the extent of the coatoms is updated consistently, and the other invocations of the *FindLowerNeighbours* function are performed using the expanded context.

After constructing the hybrid concept lattice, CREDO presents the user with the results. CREDO's interface is illustrated in Figure 4.10 with the query 'jaguar'. The top frame contains a search box with the query submitted, the lower left-hand frame shows the lattice top along with its most numerous children using a hierarchical folder representation, and the lower right-hand frame shows the documents associated with the currently selected concept (at

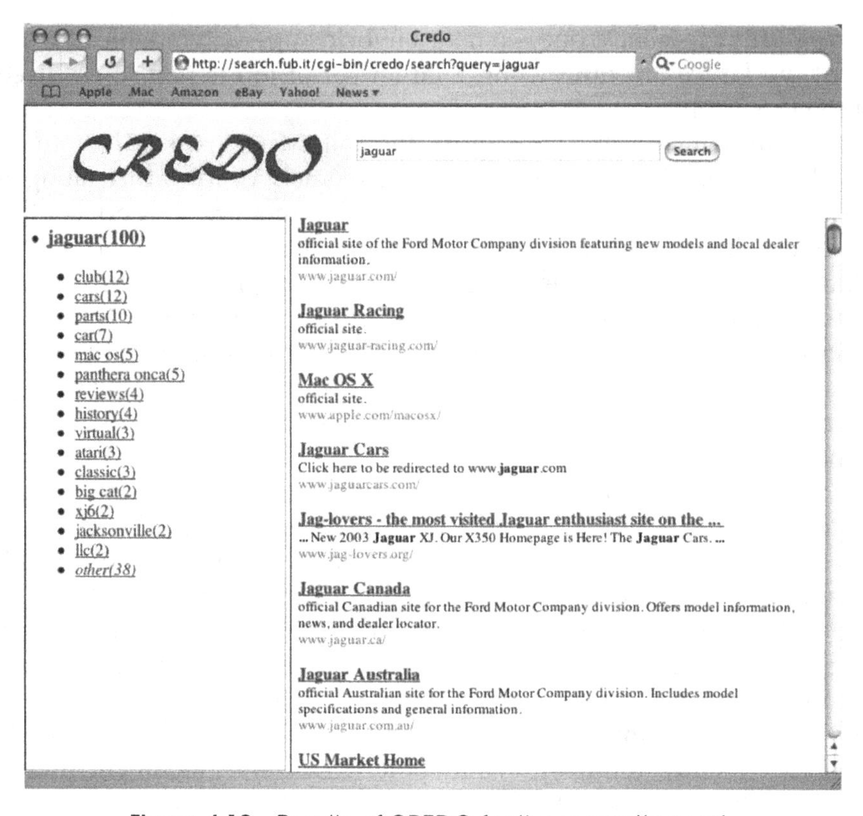

Figure 4.10 Results of CREDO for the query 'jaguar'.

the outset, the top level of the lattice). All the documents of one concept that are not covered by its displayed children are grouped into a dummy concept named 'other' (in Figure 4.10, 'other' contains 38 out of the 100 documents associated with its parent).

Note that, as already pointed out in the previous case study, the sum of the documents covered by the children may exceed the number of documents contained in their parent, because the same document can be assigned to multiple children. In Figure 4.10, for instance, the top element of the lattice contains the 100 documents returned by the search engine, whereas the sum of the documents contained by its children is equal to 114.

The user can click on one concept and see its children, thus narrowing down the scope of the search. This operation can be repeated on the newly displayed concepts to further narrow the scope of the search, or it can be performed on other top concepts to browse unexplored branches.

To manage the interaction between the user and CREDO through a Web connection, different technical solutions are conceivable. One possibility is to feed the client with just the top level and then provide the lower levels on demand, but this increases the number of interactions between client and server and requires proper handling of the *state* on the part of the server. An alternative approach is to compute the whole lattice and send out the results of all possible concept selections at once. This requires more bandwidth but is simpler to implement and allows faster response times during the interaction. CREDO adopts the latter solution.

As CREDO allows the user to browse through the query results, this can be seen as a form of query refinement, discussed in Section 3.1. In fact, there are major differences.

Rather than building a large concept lattice for an entire collection off-line, mapping a user query onto the constructed lattice, and showing a small portion of it for refining the query, CREDO constructs a small concept lattice from the retrieved documents on the fly, and shows it in full to the user for browsing through the results.

In CREDO, the emphasis is thus on the efficient construction of a small but informative concept lattice suitable for content browsing.

Furthermore, we envisage CREDO more as a tool for query disambiguation and content categorization than for strict document retrieval. This is especially useful for Web searches, because as there are a huge number of highly rich and

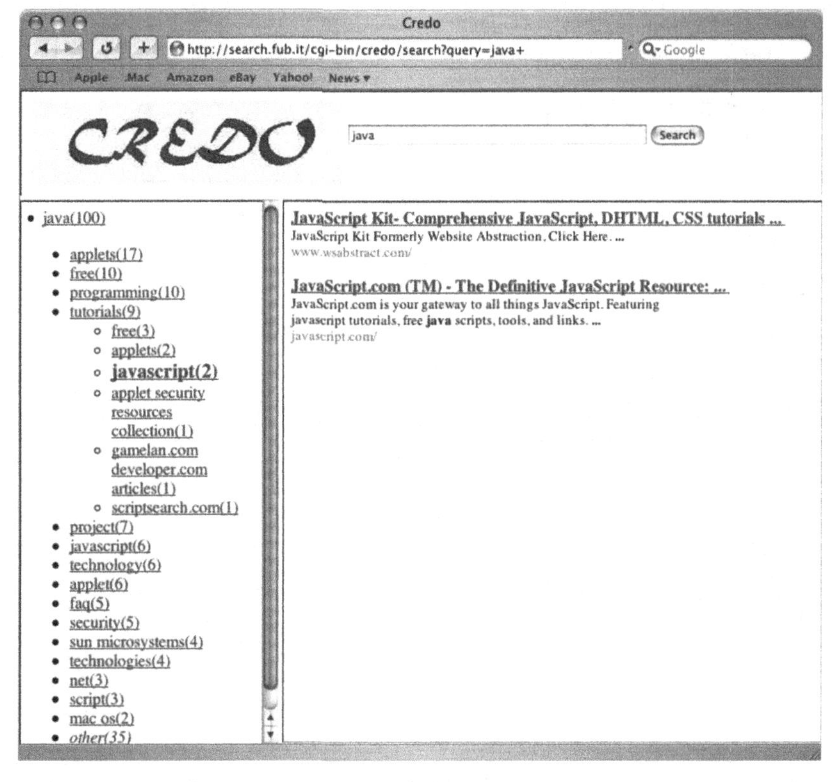

Figure 4.11 Selecting *tutorials* and then *javascript* in the 'java' results.

heterogeneous documents, the retrieval results may easily contain potentially relevant documents that cover entirely different subjects. This point is clearly illustrated in the next section.

4.2.3 Example sessions

Consider again the query 'jaguar'. This is an inherently ambiguous word on the Web. The output of CREDO (see Figure 4.10) clearly reflects this fact, with concepts such as *cars*, *parts*, *atari*, and *xj6* (referring to the Jaguar car brand), *mac os* (one of the latest operating systems of MAC), *panthera onca* and *big cat* (i.e., the animal), *jacksonville* (i.e., the American football team), and even *clubs* (there are in fact plenty of clubs called jaguar).

This example shows the utility of CREDO for disambiguating a user query and quickly focusing on the documents relevant to the intended meaning.

As a further illustration, consider the query 'java'. The great majority of the documents retrieved are about the Java programming language, as shown in Figure 4.11. The coatoms extracted from CREDO show several useful main topics referring to it, such as *applets*, *tutorials*, *javascript*, *security*, *faq*, *sun*

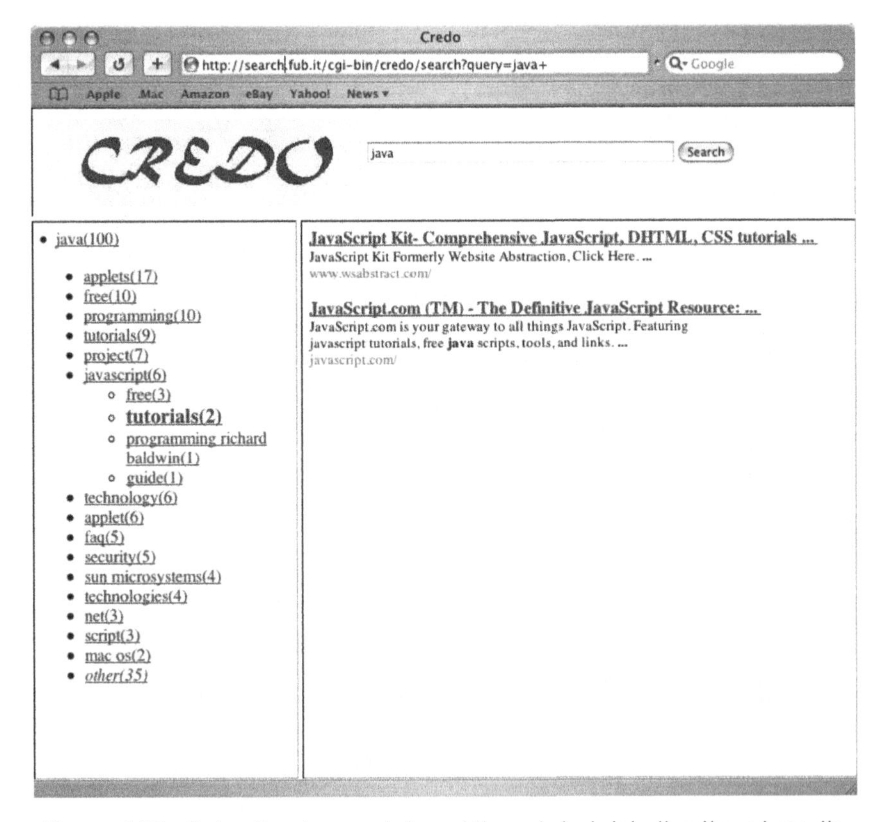

Figure 4.12 Selecting *javascript* and then *tutorials* in the 'java' results.

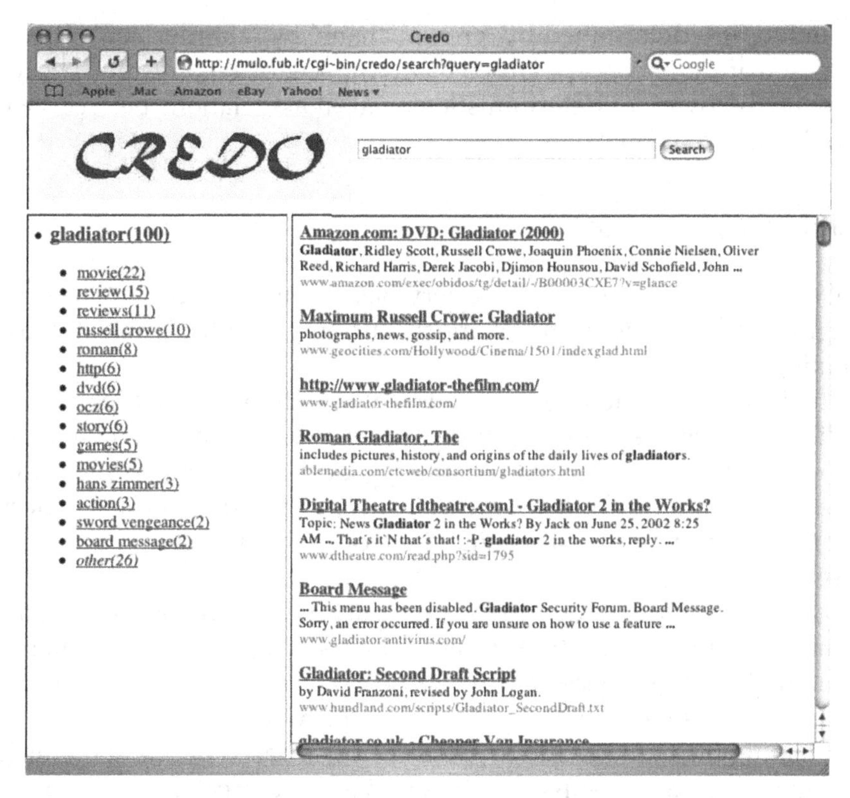

Figure 4.13 Results of the query 'gladiator'.

microsystems, etc. An example of secondary topic is 'tutorials on Javascript', for which relevant documents can be found by choosing *tutorials*, and then *javascript* (as in Figure 4.11), or by looking at *javascript*, and then *tutorials* (see Figure 4.12). Neither path is better than the other, but one may better fit a particular user's paradigm or need.

This example illustrates the greater flexibility of navigating CREDO's results, as compared to using a strict hierarchy, because the same piece of information can be reached through multiple paths.

The last example concerns the query 'gladiator', shown in Figure 4.13. There emerge many interesting concepts, mostly referring to the famous movie *Gladiator*, such as *movie*, *review* and *reviews*, *russell crowe* (i.e., the star of the movie), *hans zimmer*, who composed the *Gladiator* soundtrack, *games* and *sword vengeance* for *Gladiator*-based games, or even *ocz* (i.e., the name of the firm producing the Gladiator heatsink).

Note that several of these concepts are multiple-word concepts (*russell crowe*, *hans zimmer*, *sword vengeance*). In fact, in the limited context represented by the results of 'gladiator', each word in any of these pairs always co-occurs with the other word in the pair. For instance, 'russell' univocally determines and is univocally determined by 'crowe', 'hans' by 'zimmer', etc. This is an interesting feature of concept lattice-based mining of Web results, because it permits the discovery of deterministic or causal associations between words that hold in a given context.

4.3 Bibliographic notes

The effective integration of the query-based mode with the navigation paradigm has been the focus of much current research on text retrieval and mining.

One typical choice is to maintain different search methods in parallel (e.g., [151], [99]); in this case, the integrated system is, in practice, like a switch whereby the user may select either strategy. A tighter form of integration is achieved by cascading the two strategies, e.g., browsing prior to querying [183], or querying prior to browsing [149], or by having them coexist in the same search space ([1], [110]).

In these forms of integration the system may have to maintain several data structures, possibly supporting different kinds of operations. Furthermore, when a single search space is used to host the two strategies consistency problems may arise. For instance, merging a browsable hierarchy of directories containing file names with virtual directories based on file contents [110] may yield inconsistent results if a file is moved to a directory which does not conform to its content.

As seen in the chapter, concept lattices take the hybrid searching paradigm one step further, because multiple retrieval strategies, including querying and navigation, can be seamlessly integrated in a unique search space. To characterize this state of things, in [38] the metaphor of the GOMS user's cognitive model [31] and user activity [169] is used.

The merits of using concept lattices to support hybrid search strategies have been described in a number of papers (e.g., [106], [39], [79], [62], [78]). They can be summarized as greater flexibility, good retrieval effectiveness, and mining capabilities. One of the most interesting application domains is file system management: Ferré and Ridoux [79] describe a conceptual file system model where concepts serve both as directories of files and as queries expressing file properties.

One important form of text mining that can be incorporated into a concept lattice, not considered in this chapter, is the combination of partial views. When the data can be classified along multiple axes (e.g., functional, geographical, descriptive), it may be convenient for the user to bring in new attributes in an incremental fashion, making decisions based on the information displayed by the system for the current choice of the attributes. This approach can be effectively implemented using the nesting and zooming technique described in Section 2.3.2, and it can be naturally combined with the other retrieval strategies considered in this chapter.

Using view combination is most suitable for mining the information contained in domains characterized by semistructured data, because it may be easier to identify valuable subcontexts. One of the most interesting applications developed is a system for searching collections of e-mails ([63], [62]), whereby the user can combine the information contained in different fields of the messages (e.g., sender and subject). A commercial follow-up of this system, inspired by Peter Eklund, has recently reached the market under the name of Mail-Sleuth (http://mail-sleuth.com), with a freely available academic software variant (http://www.hiermail.com). Analysing real-estate data extracted from the Web [61] is in the same line of research.

Turning to the rendering of retrieval results, the need for concise display and user-oriented manipulation was recognized before the advent of Web-based search services. Among various other systems, BEAD [50] and LyberWorld [124] depict clustering patterns using three-dimensional visualization schemes, InfoCrystal [219] is based on a particular visual representation of a Venn diagram to suggest how to refine Boolean queries, and TileBars [121] uses a tile mosaic metaphor to show which parts of each document match the query terms. The approach taken by Veerasamy and Heikes [246] should also be mentioned, as they adopt a histogram metaphor to display the weight of each query term in each document, as should the work of Byrd [30], who exploits the scrollbar to show where terms are within each individual document. A comprehensive treatment of the visualization issue in information retrieval is given by Hearst [122].

Recent research into the visualization of retrieval results has focused on the lack of effectiveness of Web search engine interfaces. A number of approaches have been presented, including query-biased summaries [238], hierarchical query-biased summaries [232], query term hits between documents [21], and arrangement of results using an auditorium seating metaphor [235].

Clustering Web results has been investigated by, among others, Zamir and Etzioni [277]. They create clusters based on strings of words shared between documents, so that cluster labelling is merged with cluster formation. Clustering of Web results is also offered by commercial search engine services such as Vivísimo (http://vivisimo.com) or Clustered Hits (http://www.clusteredhits.com).

The use of distributional scoring functions for selecting the most relevant terms has been proposed for solving various retrieval tasks, although not directly related to cluster-based browsing of Web results (e.g., [120], [197], [14]). The Kullback–Leibler distance is well known in information theory ([67], [143]); it has been used as a theoretical basis for retrieval feedback by Carpineto *et al.* [33]. The combination of several distributional term-scoring functions, including the Kullback–Leibler distance, to improve the quality of the generated terms is discussed in [47].

Clustering of Web retrieval results is a particular form of Web mining. The field of Web mining covers several other aspects not addressed in this book, such as Web crawling and link-based page ranking. These issues are discussed, for instance, in [49] and [211].

5

Rule Mining

Rule mining is the process of finding dependencies across data. Thanks to its mathematical properties, a concept lattice can be used to make rule mining more systematic, or faster, or more accurate. In this chapter we will use the concept lattice built from a context as an intermediate support structure to mine multiple data dependencies that hold in the context. We consider both deterministic rules, which always hold across the data, and probabilistic rules, which may not always hold. In particular, in the four sections of the chapter the following rule mining tasks will be considered: (i) inference of implications, (ii) inference of functional dependencies, (iii) extraction of association rules, (iv) induction of classification rules.

5.1 Implications

We start by introducing the definition of implications (or implication rules). Without loss of generality, we refer to a single-valued context. If the original context is many-valued, we need to transform it into a single-valued one, as discussed in Section 1.3.

A context (G, M, I) satisfies the implication $Q \rightarrow R$, with $Q, R \subseteq M$, if

for all $g \in G, gIq$ for all $q \in Q$ implies gIr for all $r \in R$.

In other words, an implication between two subsets of attributes Q and R means that if a set of objects is described by the attributes contained in Q then it is necessarily described by the attributes contained in R. In particular, an implication $Q \rightarrow R$ holds when there is no object that contains all the attributes in Q (we say that the implication holds vacuously), although in many circumstances it may be convenient to require at least one supporting object.

We will use as an illustration throughout the chapter a slightly more elaborate version of the planets context introduced in Table 1.2, taken

Concept Data Analysis: Theory and Applications. Claudio Carpineto and Giovanni Romano
 ISBN: 0-470-85055-8

Table 5.1 An enlarged context for the planets, including 'period'

	size			*distance from sun*		*moon*		*period*	
	small	medium	large	near	far	yes	no	short	long
Mercury	×			×			×	×	
Venus	×			×			×	×	
Earth	×			×		×		×	
Mars	×			×		×		×	
Jupiter			×		×	×		×	
Saturn			×		×	×		×	
Uranus		×			×	×			×
Neptune		×			×	×			×
Pluto	×				×	×			×

from [46]. The new context, characterized by the presence of the additional attribute 'period' (short/long), is shown in Table 5.1. As explained in Section 1.3, this many-valued context can be seen as a single-valued context in which each object is described by one attribute–value pair per many-valued attribute.

There are many implications that hold in the enlarged planet context. Some examples of implications are: $dn \rightarrow ss$, i.e., a planet which is near the sun has a small size; $dn \rightarrow ps$, i.e., a planet which is near the sun has a short period; $ss \;\; ps \rightarrow dn$, i.e., a planet of small size with a short period is near the sun; and $sl \;\; mn \rightarrow df$, i.e., a planet of large size without a moon is distant from the sun. Note that the last implication holds vacuously, because there is no planet of large size which does not have a moon.

Although it is always theoretically possible to find all the implications that a context satisfies, because there is only a finite number of subsets of M, such an approach would be impractical. Furthermore, the complete set of implications will, in general, contain several trivial or redundant elements.

It is thus more convenient to find a restricted set of informative implications from which the other implications can be derived by logical implication. For this purpose, we recall the following concepts, borrowed from database theory.

Given a set Σ of implications, the *closure* Σ^+ is the set of implications derived from Σ by a *complete* set of inference axioms. By complete we mean that every implication that is logically implied by Σ can be derived from the implications in Σ by one or more applications of these axioms.

This issue has been thoroughly studied by Maier [152]. One complete set of axioms is the following:

1. Reflexivity: $Q \rightarrow Q$.
2. Augmentation: $Q \rightarrow R$ implies $Q \cup Z \rightarrow R$.
3. Additivity: $Q \rightarrow R$ and $Q \rightarrow Z$ implies $Q \rightarrow R \cup Z$.
4. Projectivity: $Q \rightarrow R \cup Z$ implies $Q \cup Z \rightarrow R$.
5. Transitivity: $Q \rightarrow R$ and $R \rightarrow Z$ implies $Q \rightarrow Z$.
6. Pseudo-transitivity: $Q \rightarrow R$ and $R \cup Z \rightarrow W$ implies $Q \cup Z \rightarrow W$.

It is also possible to consider only the first, second, and sixth axioms, from which the others can be proven. These are sometimes called Armstrong's axioms.

Two sets Σ and Ω are said to be *equivalent* if they have the same closure. If Σ and Ω are equivalent, then Ω is called a *cover* for Σ. This definition of cover makes no mention of the relative sizes of Σ and Ω, although we are usually interested in finding restricted covers.

Restricted covers can be defined in several ways depending on which aspect we would like to emphasize (e.g., logical redundancy, minimality of the antecedents, number of implications, global number of attribute symbols required, etc.).

One possible choice is to search for a *non-redundant* cover. A cover Ω is non-redundant if there is no proper subset of Ω that is equivalent to Ω itself. This amounts to saying that a non-redundant cover contains only implications that do not follow from the remaining implications. Non-redundant covers are difficult to generate because it is usually required that some redundant cover is found first, from which redundant elements are then removed. Note that a set of implications can contain more than one non-redundant cover.

An alternative characterization of a non-redundant cover relies on the notion of *pseudo-intent*. A set $P \subseteq M$ is called a pseudo-intent of (G, M, I) if and only if

$$P \neq P'' \text{ and } Q'' \subseteq P \text{ holds for every pseudo-intent } Q \subseteq P, Q \neq P$$

(the doubleprime operator is the same as used in Section 1.2).

The set of all implications of the form $P \rightarrow (P'' \backslash P)$, called a Duquenne–Guigues basis, is a non-redundant cover [115]. A direct utilization of this recursive definition is theoretically possible because M is finite, but it would be seriously inefficient, as the number of pseudo-intents may grow exponentially and pseudo-intents may be difficult to assess.

Apart from computational problems, non-redundant covers may present practical limitations, because it may be difficult for the user to *logically* derive (e.g., by transitivity) implications of interest not explicitly contained in the cover.

If a cover is non-redundant, it is guaranteed that it cannot be made smaller by removing implications. However, it may still be possible to remove attributes from its implications. This leads to the notion of reduced cover.

A cover Ω is *left-reduced* if there is no $Q \rightarrow R$ in Ω and no proper subset S of Q such that the set $\Omega - \{Q \rightarrow R\} \cup \{S \rightarrow R\}$ is equivalent to Ω. A cover is *right-reduced* if the corresponding statement holds for the right-hand side of each implication. A cover is *reduced* if it is left-reduced and right-reduced. Reduced implications are more readable, although they are not necessarily the most useful implications.

Note that a reduced cover may well be redundant, just as a non-redundant cover may not be reduced. A reduced cover is sometimes called a most (or maximally) general cover; in particular, it may be required that every consequent contains only one attribute.

Before addressing the problem of the automatic determination of a specific type of cover for the set of implications that hold in a context, we wish to characterize their size. The computational space complexity of implications is both a theoretically interesting problem and a key factor for applications.

5.1.1 Computational space complexity of implications

First of all, it should be noted that there are some theoretical upper bounds on the number of holding implications that can be expressed as a function of the three main input parameters of the problem: the number of objects $|G|$, the number of attributes $|M|$, and the number of values per attribute v.

As there are $|M|$ attributes with v values each, the conjunctive language used to describe the implications will contain $[(v+1)^{|M|-1} - 1]\ |M|v$ possible distinct elements, regardless of the number of objects. On the other hand, with $|G|$ objects it is possible to generate at most $(2^{|M|-1} - 1)\ |M|\ |G|$ implications, i.e., when the $|G|$ objects have no attribute–value pairs in common. The actual theoretical bound is the smaller of the two.

While it is easy to show some context in which the set of implications does grow exponentially with respect to the number of attributes even without considering vacuous implications (e.g., it suffices to consider a context where all the objects are described by the same attribute value), this situation should be rare in practice.

To analyse the problem on a more quantitative basis, we derive a formula for the expected number of implications (*NI*) that hold in a database under the assumption of uniform distribution of the attribute values. We denote the left-hand side (or antecedent) of an implication by *lhs*, and the right-hand side (or consequent) by *rhs*. For simplicity, we assume without loss of generality that every consequent contains only one attribute (i.e., $|rhs| = 1$).

The formula can be obtained by summing the probabilities of occurrence of each possible implication. To make this computation feasible, it is convenient to consider the set of implications with the same *rhs* and with $|lhs| = k$

$(1 \le k \le |M| - 1)$, where each implication has the same probability of occurrence. There are $\binom{|M|-1}{k} v^k$ possible distinct implications of this kind with $|lhs| = k$. The probability that there is an implication with exactly i supporting objects ($1 \le i \le |G|$) is given by the product of three factors. The first factor is the probability that the first i objects support the implication, i.e., $\left(\frac{1}{v^{k+1}}\right)^i$. The second is the probability that the remaining $|G| - i$ objects neither support nor contradict the implication, i.e., $\left(1 - \frac{1}{v^k}\right)^{|G|-i}$. This factor ensures that there is no other object supporting the implication and that the implication is not disconfirmed by some of the remaining objects. The third is the number of possible ways to choose the supporting i objects among the $|G|$ objects in the relation, i.e., $\binom{|G|}{i}$.

If we require that each implication is supported by *at least* one object, the expected number of implications with $|lhs| = k$ is

$$\binom{|M|-1}{k} v^k \sum_{i=1}^{|G|} \binom{|G|}{i} \left(\frac{1}{v^{k+1}}\right)^i \left(1 - \frac{1}{v^k}\right)^{|G|-i}. \tag{5.1}$$

To derive the complete formula we have to sum equation (5.1) over k and multiply the result by the number of possible reduced right-hand sides (i.e., $|M|\ v$):

$$NI = |M|\ v \sum_{k=1}^{|M|-1} \binom{|M|-1}{k} v^k \sum_{i=1}^{|G|} \binom{|G|}{i} \left(\frac{1}{v^{k+1}}\right)^i \left(1 - \frac{1}{v^k}\right)^{|G|-i}. \tag{5.2}$$

Equation (5.2) shows that NI grows exponentially with the number of attributes, similar to the theoretical upper bound. Equation (5.2) also shows that NI grows monotonically with v, from $(2^{|M|-1} - 1)\ |M|$, for $v = 1$ (i.e., all possible implications associated with one distinct object), to $(2^{|M|-1} - 1)\ |M|\ |G|$ as v tends to infinity (i.e., the theoretical upper bound, holding when the objects have no attribute–value pairs in common).

Equation (5.2) can be written more compactly as:

$$NI = |M|\ v \sum_{k=1}^{|M|-1} \binom{|M|-1}{k} v^k \left[\left(1 - \frac{1}{v^k} + \frac{1}{v^{k+1}}\right)^{|G|} - \left(1 - \frac{1}{v^k}\right)^{|G|}\right]. \tag{5.3}$$

Figure 5.1, taken from [45], shows results plotted from equation (5.3) for four pairs of values for $(|M|, v)$: (5, 5), (10, 2), (10, 5), (10, 10). For $n = 1$, $NI = |M|\ 2^{|M|-1}$; i.e., for each attribute–value pair contained in the object it is

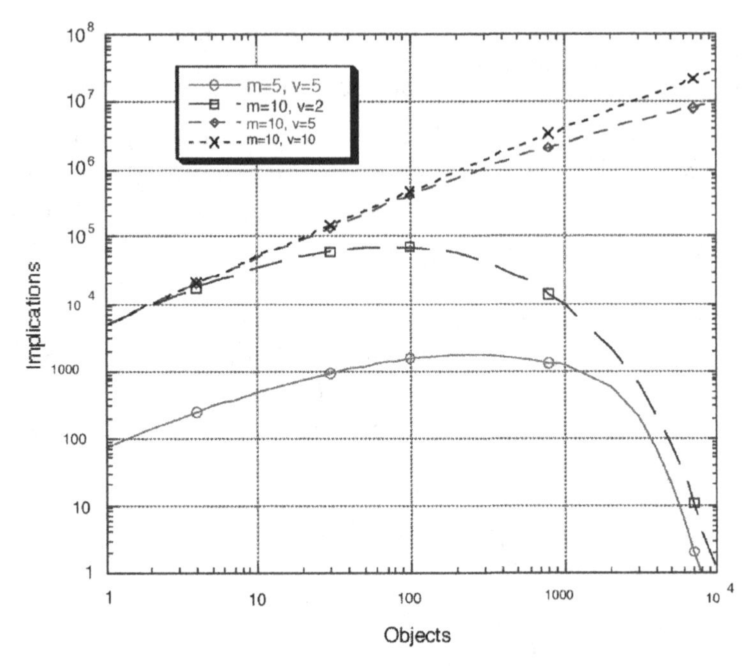

Figure 5.1 Theoretical size complexity of implications under uniform distribution. Both scales are logarithmic.

possible to form implications with all possible combinations of the remaining $|M|-1$ attribute–value pairs. For each fixed pair $(|M|, v)$, as long as $n \ll v\,|M|$, *NI* grows linearly with respect to $|G|$. For larger values of $|G|$, since previously unseen attribute–value pairs are encountered less frequently, *NI* decreases; in particular, as $|G|$ grows to infinity, *NI* tends to zero because when all possible combinations of attribute–value pairs have been seen there exist no more implications in the database. This behaviour is apparent for the two lower curves in Figure 5.1, in which the object-description space is small (55 and 210, respectively). Finally, Figure 5.1 shows that, for a fixed value of $|M|$, as v grows the value of *NI* remains nearly stable, provided that $n \ll v\,|M|$ (see the two upper curves in Figure 5.1).

We must emphasize that these findings hold for random descriptor assignment, which may not be the best model for the distribution observed in many practical situations. Indeed, one might expect that the number of implications in a natural data set should be much smaller, because this type of rule does not admit exceptions.

On the other hand, a number of experimental results confirm that the size of the set of implications that hold in a *real* context can grow exponentially with respect to the number of attributes, with data sets of small size (up to thousands of objects and tens of attributes) yielding tens of thousands of non-vacuous implications [45].

5.1.2 Generating implications from the concept lattice

In order to mine the implications that hold in a context, we can take advantage of the concept lattice of the context. The fundamental observation that allows us to read off an implication from the concept lattice is the following. An implication $Q \to R$ holds in a context (G, M, I) if and only if

$$(R', R'') \geq (Q', Q''),$$

i.e., if and only if

$$\bigwedge \{\mu(r) \mid r \in R\} \;\geq\; \bigwedge \{\mu(q) \mid q \in Q\}.$$

In other terms, $Q \to R$ holds if and only if the largest concept containing Q as a part of its intent is also described by R. Note that we may equivalently verify whether the same condition is separately fulfilled by all implications $Q \to r$, with $r \in R$.

Returning to our illustrative context in Table 5.1, we show in Figure 5.2 the concept lattice derived from it. It is evident, for instance, that the implication $dn \to (ss\ ps)$ holds in the context, because the largest concept containing dn in the corresponding concept lattice also contains ps and ss, whereas the implication $my \to df$ does not hold, because the largest concept containing my does not contain df.

A consequence of the proposition stated above is that the implications that can be generated from a concept (X, Y) are of the form $Q \to R$, with $Q \cap R = \varnothing$, $Q \subseteq Y$, $R = Y \backslash Q$, provided that there is no parent (W, Z) of (X, Y) such that $Q \subseteq Z$. If the last condition is satisfied, we say that Q is *consistent* with the parents of (X, Y). It also follows that in order to find the complete set of implications it is sufficient to collect all the implications generated from each concept (the generation of implications that hold vacuously can be avoided by omitting to consider the bottom element of the lattice).

Such a set, however, will in general be highly redundant. One source of redundancy is represented by the fact that in the set of implications associated with each concept there may be implications that can be obtained from some other implication r in the same set by shifting attributes from the right-hand side of r to its left-hand side; i.e., by projectivity and by augmentation, in terms of Maier's axioms.

For example, the concept with intent $\{sl\ df\ my\ ps\}$ in Figure 5.2 yields the implication $sl \to (df\ my\ ps)$, but it also generates the implications $(sl\ df) \to (my\ ps)$, $(sl\ my) \to (df\ ps)$, $(sl\ ps) \to (df\ my)$, and $(sl\ df\ ps) \to (my)$, all of which can be derived from the former by attribute shifting.

To cope with this kind of redundancy, it is convenient to make use of the partial order on $\wp(Y)$ given by standard set inclusion. We will say that element c_1 is more general than c_2 if $c_1 \subset c_2$. It immediately follows that the

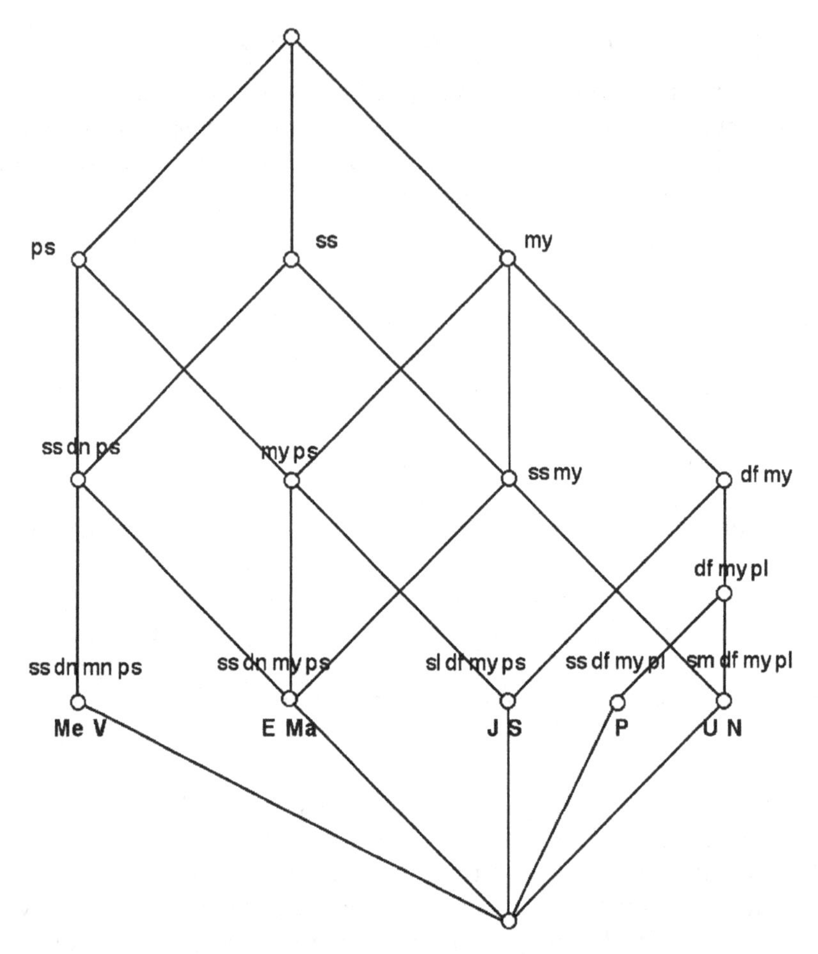

Figure 5.2 Concept lattice for the context in Table 5.1.

set of implications associated with the concept (X, Y) that are non-redundant with respect to attribute shifting are of the form $lhs \rightarrow rhs$, where lhs is a most general element of $\wp(Y)$ that is consistent with the parents of (X, Y), and rhs is given by $Y \backslash lhs$.

We now describe an algorithm for inferring an implication cover that does not contain redundant implications with respect to attribute shifting from the concept lattice of the context. Different approaches for generating restricted implication covers are discussed in Section 5.5.

For each concept (X, Y), it is necessary to find all the most general subsets of Y that are consistent with the parents of the concept. The algorithm takes advantage of the partial ordering relation defined by generality over the elements of $\wp(Y)$, because the resulting ordered set is well formed with respect to the task at hand. In particular, if an element of $\wp(Y)$ is consistent (with the parents of the concept), then all its more specific elements are also consistent; conversely, if an element is inconsistent then all its more general elements are also inconsistent.

The algorithm examines one parent at a time, maintaining and updating a set *lhsSet* containing all the most general elements that are consistent with the parents that have been seen so far. *lhsSet* acts as an upper boundary for the hypothesis space: more specific hypotheses are consistent while more general hypotheses are inconsistent and do not need be considered.

lhsSet is updated in the following way. Initially, it contains the most general elements of $\wp(Y)$, i.e., those formed by single attributes. Then, for each parent of (X, Y), the procedure deletes all the elements that are inconsistent with the parent from *lhsSet*, and replaces each with its most general consistent specializations (a specialization of an element *lhs* in *lhsSet* is any element of $\wp(Y)$ that is more specific than *lhs*). Finally, it removes the consistent specializations that are more specific than (or equal to) some other elements of *lhsSet*.

The key operation is the determination of the most general consistent specializations. While finding all the most general specializations of a given element is straightforward (i.e., it is sufficient to add a single attribute to the element in all possible ways), a potentially demanding aspect is the number of minimal specialization steps that must be performed before each specialization becomes consistent.

Fortunately, it turns out that all most general consistent specializations of an element *lhs* of *lhsSet* belong to the set of the most general specializations of *lhs*. This result follows from the observation that (i) in the set of the most general specializations of the elements of *lhsSet* there is at least one consistent specialization (unless the concept (X, Y) generates no implication at all), and (ii) more specific consistent elements of $\wp(Y)$ are supersets of some consistent element contained in the set of the most general specializations of *lhsSet*.

Therefore, for each element *lhs* that needs to be specialized, it is sufficient to generate and test only the most general specializations of *lhs*. As more and more parents are examined, the boundary *lhsSet* moves down and the hypothesis space shrinks, until it may become empty if the concept yields no implication at all.

A detailed description of the algorithm is given in Figure 5.3. We again assume without loss of generality that the input context is a single-valued context.

To illustrate, consider the determination of the set of implications associated with the concept with intent (*ss df my pl*) in the concept lattice shown in Figure 5.2. In Figure 5.4 we trace the working of the *Find Implications* algorithm. The *lhsSet* computed for the concept at hand is {(*ss df*), (*ss pl*)}; the corresponding implications are (*ss df*) $\rightarrow$ (*my pl*) and (*ss pl*) $\rightarrow$ (*df my*).

```
Find Implications
Input: A concept lattice L of context (G, M, I)
Output: A cover contextImpls for the implications that hold in (G, M, I)
1. contextImpls := ∅
2. for each (X, Y) ∈ L
3. contextImpls := contextImpls ∪ FindImplicationsFromConcept((X, Y), L)

function FindImplicationsFromConcept((X, Y), L)
/* Returns the set of implications derivable from concept (X, Y) */
1. lhsSet := Y
   /* Contains the left-hand sides of the implications derivable from concept */
2. for each (W, Z) parent of (X, Y)
3.    updLhsSet := ∅    /* Contains the elements of lhsSet updated by parent */
4.    for each lhs ∈ lhsSet
5.       if lhs ⊈ Z
            then
6.             updLhsSet := updLhsSet ∪{lhs}
            else
7.             candSpecSet := ∅
               /* Contains candidate specializations of the elements in lhsSet */
8.             for each m ∈ Y\ lhs
9.                candSpecSet := candSpecSet ∪{lhs ∪{m}}
10.            for each candSpec ∈ candSpecSet
11.               if ConfirmSpec(candSpec, lhs, Z, lhsSet) = true then
12.                  updLhsSet := updLhsSet ∪{candSpec}
13.   lhsSet := updLhsSet
14.conceptImpls := ∅    /* Contains all implications derivable from concept */
15.for each lhs ∈ lhsSet
16.   conceptImpls := conceptImpls ∪{lhs → Y\lhs}
17.   return conceptImpls

function ConfirmSpec(candSpec, lhs, Z, lhsSet)
/* Checks if a spec. is consistent with a parent and more general than current specs */
1. if candSpec ⊈ Z and there is no element ⊆ candSpec in lhsSet\{lhs} then
2.    return true
```

Figure 5.3 The Find Implications algorithm for inferring a cover for the implications that hold in a context from the concept lattice of the context.

The complete set of implications returned by the algorithm (ignoring the bottom element of the lattice) is as follows:

$$
\begin{aligned}
pl &\rightarrow df\ my \\
df &\rightarrow my \\
ss\ ps &\rightarrow dn \\
dn &\rightarrow ss\ ps \\
mn &\rightarrow ss\ dn\ ps \\
dn\ my &\rightarrow ss\ ps
\end{aligned}
$$

$$\begin{aligned}
ss\ my\ ps &\rightarrow dn\\
df\ ps &\rightarrow sl\ my\\
sl &\rightarrow df\ my\ ps\\
sm &\rightarrow df\ my\ pl\\
ss\ pl &\rightarrow df\ my\\
ss\ df &\rightarrow my\ pl
\end{aligned}$$

The computational time complexity of the Find Implications algorithm is as follows. The main loop of the algorithm iterates on the concepts in the lattice. Each iteration invokes the *FindImplicationsFromConcept* function. For each parent of the concept, the updating of *lhsSet* requires time at most proportional to $k^2|M|$, where k is the largest size of *lhsSet* and $|M|$ is the number of attributes. Thus, the time complexity of the Find Implications algorithm is at most proportional to $O(|C|k^2|M|q)$, where $|C|$ is the number of concepts and q is the largest number of parents per concept.

```
lhsSet = {ss, df, my, pl}
parent (ssmy)
    updLhsSet:= ∅
    lhs := ss
        candSpecSet := {(ss df), (ss my), (ss pl)}
        lhsSet := ∅    /* Current specializations fail confirmation test */
    lhs := df
        updLhsSet := {df}
        lhsSet := {df}
    lhs := my
        candSpecSet := {(my ss), (my df), (my pl)}
        lhsSet := {df}    /* Current specializations fail confirmation test */
    lhs := pl
        updLhsSet := {df, pl}
        lhsSet := {df, pl}
parent (dfmysl)
    updLhsSet:= ∅
    lhs := df    /* Loop for element in current lhsSet */
        candSpecSet := {(df ss), (df my), (df pl)}
        updLhsSet := {(df ss)} /* Only one current spec. passes conf. test */
        lhsSet := {(df ss)}
    lhs := pl
        candSpecSet := {(pl ss), (pl df), (pl my)}
        updLhsSet := {(df ss), (pl ss)} /* Only one current spec. passes conf. test */
        lhsSet := {(df ss), (pl ss)}
conceptImpls = {(ss df) → (my pl), (ss pl) → (df my)}
```

Figure 5.4 Step-by-step computation of the implications derivable from the concept with intent (*ss df my pl*) in the concept lattice in Figure 5.2 using the Find Implications algorithm.

The cover returned by the algorithm will, in general, contain many redundant implications, as in this case. For instance, the set of implications shown above contains both $dn \rightarrow ps$ and $(my\ dn) \rightarrow ps$. On the other hand, all the implications returned by the algorithm, including the redundant ones, are guaranteed to hold non-vacuously.

If we are interested in a non-redundant cover, in the sense of the definition provided above, we need to apply an algorithm for removing the redundant implications from the given cover. As the removal of redundant implications requires checking whether each implication is implied by those remaining, the time complexity of this problem is typically polynomial with respect to the number of implications. A detailed description of an algorithm for removing redundant functional dependencies, which can be easily adapted to implications, is given, for instance, on page 73 of [152].

A non-redundant cover for the set of implications given above is:

$$
\begin{aligned}
pl &\rightarrow df \\
df &\rightarrow my \\
ss\ ps &\rightarrow dn \\
dn &\rightarrow ps \\
dn &\rightarrow ss \\
mn &\rightarrow dn \\
df\ ps &\rightarrow sl \\
sl &\rightarrow ps \\
sl &\rightarrow df \\
sm &\rightarrow pl \\
ss\ df &\rightarrow pl
\end{aligned}
$$

Before concluding this section, we briefly address the scope of the rules that will be considered throughout this chapter, including implications.

The rules considered are intended for categorical (or nominal) attributes, with the extension to numeric attributes being difficult at best. The common approach consists of partitioning the entire range of values of each numeric attribute into a set of intervals which can then be mapped onto a set of integers, thus creating a set of categorical values for that attribute. The optimal number of intervals depends on the characteristics of the data set; usually, it represents a compromise between using a strong discretization (e.g., binary attributes), which would compress the diversity of the objects, and using a weak discretization, which hides their similarity.

An additional difficulty in many real applications is the issue of missing values. To deal with this problem, we can try to 'guess' the missing attribute values, for instance by using the overall modal (most frequent) value for nominal attributes and Boolean features and the overall mean for numeric attributes. Alternatively, we could just ignore them; this would amount to assigning to each object with a missing attribute value a unique, unmatchable value for that attribute.

5.2 Functional dependencies

A classical form of data dependency between the attributes present in a many-valued context is functional dependency. A many-valued context (G, M, V, I) satisfies the functional dependency $H \to K$, with $H, K \subseteq M$, if for all $g_1, g_2 \in G$,

$$h(g_1) = h(g_2) \text{ for all } h \in H \text{ implies } k(g_1) = k(g_2) \text{ for all } k \in K.$$

In other words, if we wish to verify that H functionally determines K (or K is functionally determined by H) then we need to check that every tuple of values of H maps to at most one tuple of values of K.

Intuitively, functional dependency specifies which attributes of a datum allow one to conclude other attributes. For instance, the functional dependency (*size distance*) $\to$ (*period*) holds in the context in Table 5.1 because the mapping from each possible value for the antecedent to the corresponding value for the consequent is as follows: a value of 'small' for *size* and 'near' for *distance* always maps onto a value of 'short'' for *period*, 'medium, far' onto 'long', and 'large, far' onto 'short'. The attribute *size*, for instance, does *not* functionally determine the attribute *moon* because a value of 'small' for the former maps onto values of 'yes' and 'no' for the latter.

Note that functional dependency is not symmetric. For instance, although *size* and *distance* functionally determine *period*, *period* does not functionally determine *size* and *distance*.

Functional dependencies and implications have strong relationships. Given the set of implications that hold in a context, in order to see whether the functional dependency $H \to K$ holds it is sufficient to verify that there is an implication between each combination of the values taken on by the attributes in H and some combination of the values of the attributes in K.

Deriving the set of implications from the set of functional dependencies is more difficult, because an implication may hold even though its functional counterpart is not satisfied. For instance, while the attribute *distance from sun* does not functionally determine the attribute *planet size*, the attribute–value pair 'distance near' implies the attribute–value pair 'size small'. Even the mere derivation of the set of implications associated with each given functional dependency is not trivial if we require that the implications do not hold vacuously. In this case, it is necessary to search the context to ascertain which values are actually taken on by the attributes involved. For instance, from the functional dependency (*size distance*) $\to$ (*period*) it is possible to derive both (*ssdn*) $\rightarrow$ (*mn*) and (*sm dn*) $\rightarrow$ (*mn*), but only the first implication holds non-vacuously in the enlarged planet context.

Functional dependencies are thus a stronger form of data dependency than implications; in a sense, they can be seen as an abstraction of implications. Even so, their number can grow exponentially with respect to the number of attributes. Mannila and Rähiä [155] have described an example of a context in which all the covers for the functional dependencies that hold in it are of exponential size in the number of attributes. Incidentally, the same example can also be used to show that all the covers for the *implications* that hold in a context may grow exponentially with respect to the number of attributes. In practice, however, this situation may occur infrequently, as only contexts with many different keys yield inherently large covers.

Functional dependencies are primarily used for database design, in particular to develop database relations in third normal form and to optimize trade-offs between retrieval speed, data redundancy and data reliability ([152], [240]). They have been also used for consistency checking [208] and analogical reasoning [70], under the name of 'determinations'. We now address the problem of their automatic determination.

One might think that one simple method for finding a cover for the functional dependencies that hold in a context would be to transform a cover for the implications that hold in the same context, determined, for instance, using the Find Implications algorithm described in the previous section. Such an approach, however, would be highly inefficient, in that it is not possible, in general, to derive a cover for functional dependencies from a cover for implications without applying the inference axioms. As an example, consider again the context shown in Table 5.1. In order to derive the functional dependency (*size distance*) $\rightarrow$ (*period*) we should ascertain that the following implications hold: (*ss dn*) $\rightarrow$ (*ps*), (*sl df*) $\rightarrow$ (*ps*), (*sm df*) $\rightarrow$ (*pl*), (*ss df*) $\rightarrow$ (*pl*). However, of these four implications, only the fourth is explicitly present in the cover found by the Find Implications algorithm (see Section 5.1).

5.2.1 Functional dependencies as implications of transformed contexts

There is a more convenient way to cast the functional dependencies inference problem as an implications inference problem. A *transformed* context CX_T can be deduced from a many-valued context CX in such a way that the implications that hold in CX_T are exactly the functional dependencies that hold in CX. The transformed context CX_T is a triple (G_T, M, I_T), where G_T is the set of all two-element subsets of G and

$$\{g_1, g_2\} I_T M \textit{ implies that } m(g_1) = m(g_2).$$

Table 5.2 The transformed context associated with the context in Table 5.1

	size	*distance from sun*	*moon*	*period*
(1, 2)	×	×		×
(1, 3)				×
(1, 5)	×			
(2, 3)			×	×
(2, 4)			×	
(2, 5)	×		×	
(3, 4)		×	×	
(3, 5)		×	×	
(4, 5)		×	×	×

Recalling the definition of functional dependency, it is straightforward to see that for $H, K \subseteq M$, K is functionally dependent on H in CX if and only if $H \to K$ is an implication in CX_T. Thus, the cover for the implications that hold in CX_T, determined by the Find Implications algorithm, can also be taken as a cover for the functional dependencies that hold in CX.

To illustrate, consider again the enlarged planets context. The corresponding transformed context, built from the distinct objects in Table 5.1 (i.e., Mercury and Venus are labelled as object 1, Earth and Mars as object 2, Jupiter and Saturn as 3, Uranus and Neptune as 4, Pluto as 5) is shown in Table 5.2.

We should emphasize that while the objects in the original context are described by multi-valued attributes (i.e., attribute–value pairs), the objects in the transformed context are described by single-valued attributes (i.e., just attributes, without values). In fact, the original context is a many-valued context, while the transformed context is a single-valued one. Figure 5.5, taken from [46], shows the concept lattice built from the transformed context in Table 5.2; the letters S, D, M and P stand for *size*, *distance from sun*, *moon* and *period*, respectively.

Once the concept lattice of the transformed context has been constructed, e.g., by generating one transformed object at a time and updating the corresponding (transformed) lattice with the Update by Local Structure algorithm described in Section 2.2.1, we can extract the implications from it. By running the Find Implications algorithm on the concept lattice shown in Figure 5.5, we get the following (redundant) set of functional dependencies for the context shown in Table 5.1:

$$\textit{size distance} \to \textit{period}$$
$$\textit{size period} \to \textit{distance}$$

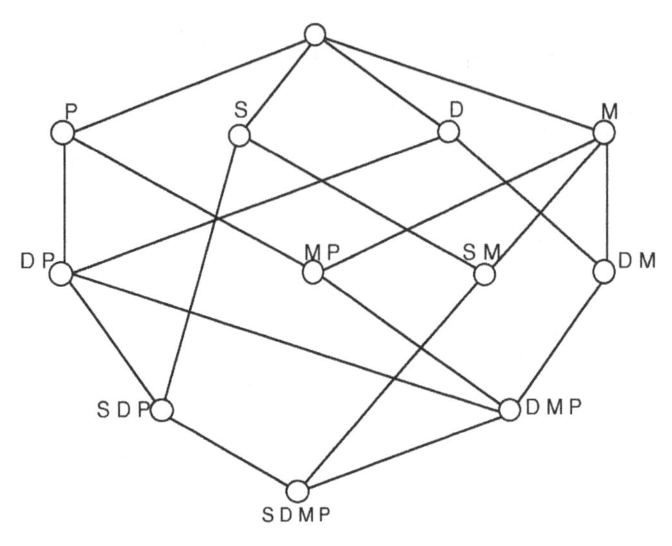

Figure 5.5 The concept lattice of the transformed context in Table 5.2.

size moon period $\rightarrow$ *distance*
size distance moon $\rightarrow$ *period*

It is worth noting that the last two functional dependencies are generated from the bottom element of the lattice in Figure 5.5. The determination of the set of functional dependencies requires that the implications yielded by the bottom element of the transformed concept lattice, which hold vacuously, should be considered.

The complexity analysis presented in Section 5.1 for the extraction of implications can also be applied to functional dependencies, provided that we consider the number of concepts $|C|_T$ present in the lattice built from the *transformed* context rather than from the original context. The algorithm can thus operate in $O(|C|_T k^2 |M| q)$ time (recall that k is the largest size of *lhsSet* and q is the largest number of parents per concept).

As the key factor for the overall time complexity is $|C|_T$, it would be useful to characterize the growth of $|C|_T$ as a function of the number of objects $|G|$. We now address this issue.

5.2.2 Computational space complexity of the concept lattice of transformed contexts

It should be noted first that $|C|_T$ grows monotonically with respect to $|G|$, because the addition of a new object increases the number of transformed objects, which, in turn, increases $|C|_T$. The growth of $|C|_T$ is not inconsistent with the fact that the number of functional dependencies that hold in a relation decreases monotonically with respect to the number of objects [206], because the addition of a new concept to a lattice usually reduces the number of implications generated by the other concepts.

A further observation is that $|C|_T$ is theoretically bounded from above by $2^{|G|-1}$, because the transformed objects, of which there are at most $\binom{|G|}{2}$, along with their possible combinations cannot exceed $\sum_{i=0}^{\lfloor |G|/2 \rfloor} \binom{|G|}{2i} = 2^{|G|-1}$. As in the untransformed case, one can use an $(n \times n)$-dimensional context containing ones in all positions but zeros along the diagonal to prove that a transformed lattice with an exponential number of concepts can actually be generated.

In practice, relations that have inherently large transformed lattices should be rare. The behaviour of $|C|_T$ can be analysed theoretically if we assume that the attribute values in the original context are uniformly distributed. In this case, each attribute in the transformed context will be assigned to each of the $|G|(|G|-1)/2$ transformed objects with (constant, independent) probability $1/v$. Under a similar hypothesis, as already seen in Section 2.1.1, the number of attributes per (transformed) object follows a binomial distribution with a mean value of $|M|/v$. In particular, the probability of occurrence of zero-attribute transformed objects will be $\binom{|M|}{0}\left(\frac{v-1}{v}\right)^{|M|}$, the probability of occurrence of one-attribute transformed objects will be $\binom{|M|}{1}\left(\frac{v-1}{v}\right)^{|M|-1}\frac{1}{v}$, etc.

The formula that gives the expected number of concepts in the transformed context, derived from equation (2.1), is

$$|C|_T = \sum_{i=0}^{\frac{|G|(|G|-1)}{2}} \sum_{j=0}^{|M|} \binom{\frac{|G|(|G|-1)}{2}}{i} \binom{|M|}{j} \left(\frac{1}{v^{ij}}\right) \left(1 - \frac{1}{v^{i}}\right)^{|M|-j} \times \left(1 - \frac{1}{v^{j}}\right)^{\frac{|G|(|G|-1)}{2} - i}. \tag{5.4}$$

Figure 5.6, taken from [46], shows results plotted from equation (5.4) for five pairs of values for $(|M|, p)$: $(7,7)$, $(20,20)$, $(20,50)$, $(50,20)$, $(50,50)$. Figure 5.1 suggests that the growth of the number of concepts in the transformed lattice varies from linear to quadratic with respect to the number of objects. The curves obtained for $(20,50)$ and $(20,20)$ represent the lower bound, whereas the curve obtained for $(50,20)$ is the upper bound.

The space complexity moves towards the upper bound as the ratio between $|M|$ and v increases; a similar behaviour holds when both $|M|$ and v grow by the same factor and the number of objects is not too small.

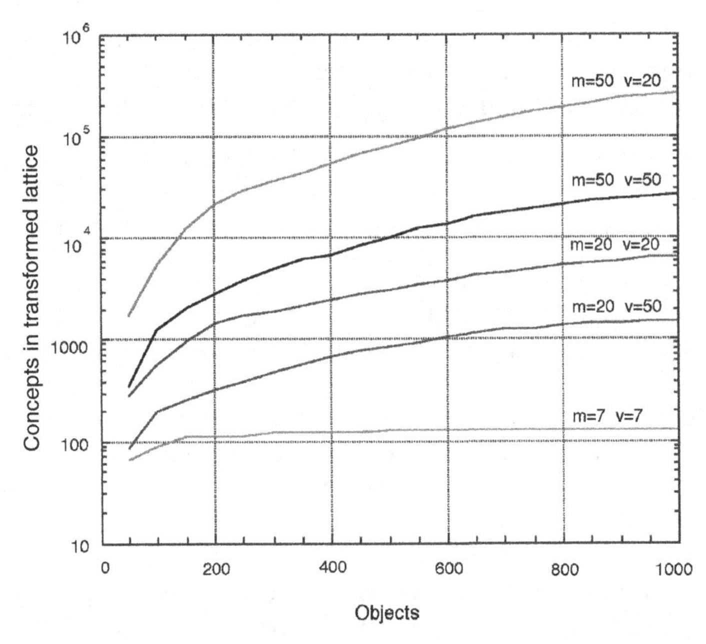

Figure 5.6 Theoretical space complexity of the concept lattice of a transformed context under uniform distribution. The vertical scale is logarithmic.

The lower curve in Figure 5.6, relating to the pair (7,7), represents a limit situation. For a fixed value of $|M|$, given that each transformed object is described by single-valued attributes, $|C|_T$ has an upper bound of $2^{|M|}$ ($2^7 = 128$ concepts in our case); hence, after a certain point the lattice is saturated (i.e., it contains 128 nodes) and $|C|_T$ remains constant as new objects are added.

It is worth noting that these results hold under the hypothesis of uniform distribution. Real databases usually produce much smaller transformed lattices, because the regularities that they exhibit typically limit the number of non-empty transformed objects and produce many identical object descriptions. By contrast, under random assignments there are a large number of distinct transformed objects, which increases the likelihood and diversity of intersections, and therefore the size of the transformed lattice. In practical situations, the size of the transformed lattice is usually comparable to, and occasionally much smaller than, the number of objects.

It is also useful to look into the relationship between the size of a normal lattice and the size of a transformed lattice for a given context. Even though one can easily construct some special context such that its transformed lattice has more concepts than its normal lattice, transformed lattices are usually much smaller than normal lattices, because transformed contexts tend to have fewer rows, fewer columns, and a greater sparseness than untransformed contexts.

5.3 Association rules

The two preceding approaches to rule mining were deterministic in nature. They yielded statements of the nature 'for all objects in the database, if an object has features a and b then it also has feature c'. In this section we consider a probabilistic approach, dealing with statements of the form 'the presence of features a and b often also involves feature c'. This approach is commonly formalized with association rules, with applications in several fields including market basket analysis, medical research, and census data.

Intuitively, an association rule $Q \to R$ holds if there are enough objects that have both Q and R and if there are enough objects among those with Q which also have R. More precisely, given $minsup \in [0, 1]$ and $minconf \in [0, 1]$, a context (G, M, I) satisfies the association rule $Q \to R_{minsup,minconf}$, with $Q, R \subseteq M$, if

$$sup(Q \to R) = \frac{|(Q \cup R)'|}{|G|} \geq minsup$$

and

$$conf(Q \to R) = \frac{|(Q \cup R)'|}{|Q'|} \geq minconf.$$

The ratios $|(Q \cup R)'|/|G|$ and $|(Q \cup R)'|/|Q'|$ are called, respectively, the *support* and the *confidence* of the rule $Q \to R$. The support is the probability of an object containing both Q and R. The confidence is the conditional probability that an object contains R, given that it contains Q. The parameters *minsup* and *minconf* are two user-supplied thresholds, for the required minimum support and minimum confidence. In this section we will usually simply write $Q \to R$, omitting the two parameters.

Theoretically speaking, association rules represent a generalization of the notion of implications, because the latter can be seen as a special case of association rules, with 100% confidence and zero or minimum support. Association rules do not, in general, support inference axioms, but they are more flexible than implications because they permit discovery of approximate dependencies.

Considering the enlarged planets context in Table 5.1, the support of the rule $df \to my$ is $5/9 = 0.55$, the support of $ss\,df \to my$ is $1/9 = 0.11$, and the support of $ss \to dn$ is $4/9 = 0.44$. The first two rules hold with 100% confidence, because there are no other objects with the same antecedent and a different consequent, whereas the rule $ss \to dn$ has a confidence of $4/5 = 0.8$.

There has recently been a great deal of work on the determination of the set of association rules that hold in a context. The problem is usually broken up into two subproblems: (i) finding all frequent subsets of attributes, also

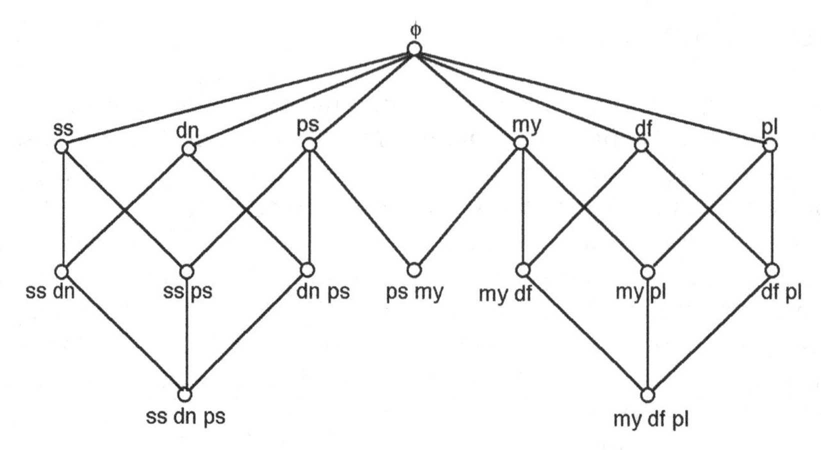

Figure 5.7 Join semilattice of frequent itemsets (*minsup* = 0.33) for the enlarged planets context.

called *frequent itemsets*, and (ii) generating confident rules from the frequent itemsets. We will illustrate both of these steps, in turn.

Figure 5.7 shows the join semilattice of all frequent itemsets with $minsup$ = 0.33 in the enlarged planets context, i.e., the itemsets contained in at least three objects. For any two itemsets, their join is always frequent while their meet may or may not be frequent. There are as many as 16 frequent itemsets, three of which ({*ss dn ps*}, {*ps my*} and {*my df pl*} are maximal, meaning that they are not a subset of any other frequent itemset. Clearly, there are also many itemsets that are not frequent; for instance, the itemsets {*ss my*} and {*dn my*} are not frequent, because they are present in only two objects, although their subsets are frequent.

The generation of the set of frequent itemsets is computationally expensive; the search space for enumeration of all frequent itemsets may grow as large as $2^{|M|}$ and for each itemset we need to compute its support, which requires checking all the objects for each attribute contained in the itemset. The situation improves if there is an upper bound l on the number of attributes describing each object; in this case the size of the search space reduces to $2^l \cdot q$, where q is the number of maximal frequent itemsets, and the computation of support may be faster.

Note that if an itemset is frequent, all its subsets are also frequent. Conversely, if an itemset is not frequent, all its supersets are also not frequent. Most frequent itemset mining algorithms are based on the last observation: $(k+1)$-itemsets (itemsets of length $k+1$) are generated only if some frequent k-itemset has been found. On the other hand, there are many application domains (e.g., telecommunications, census data, bioinformatics) where the longest frequent itemsets may easily contain tens of attributes, which makes such a method infeasible.

{*ss dn ps*}:	$ss \to dn\ ps$, $dn \to^* ss\ ps$, $ps \to ss\ dn$, $ss\ dn \to^* ps$, $ss\ ps \to^* dn$, $dn\ ps \to^* ss$
{*ss dn*}:	$ss \to dn$, $dn \to^* ss$
{*ss ps*}:	$ss \to ps$, $ps \to ss$
{*dn ps*}:	$dn \to^* ps$, $ps \to dn$
{*ps my*}:	$ps \to my$
{*df my pl*}:	$pl \to^* df\ my$, $my\ pl \to^* df$, $df\ pl \to^* my$
{*df my*}:	$my \to df$, $df \to^* my$
{*my pl*}:	$pl \to^* my$
{*df pl*}:	$pl \to^* df$

Figure 5.8 The set of association rules with *minsup* = 0.33 and *minconf* = 0.66 extracted from the frequent itemsets shown in Figure 5.7. An * indicates that the rule holds with 100% confidence.

We now turn to the generation of confident rules from the frequent itemsets. Each frequent itemset W will yield rules of the form $Z \to W \backslash Z$, with $Z \subset W$, $Z \neq \varnothing$. The rules generated from a frequent itemset will have equal support (i.e., the same support as the itemset from which they have been derived) and will, in general, have different confidence. For instance, from the frequent itemset {*ss dn ps*} we can generate the following rules, with the specified confidence (all rules have a support of $4/9 = 0.44$): $ss \to dn\ ps$ (confidence $= 4/5 = 0.8$), $dn \to ss\ ps$ (confidence $= 4/4 = 1$), $ps \to ss\ dn$ (confidence $= 4/6 = 0.66$), $ss\ dn \to ps$ (confidence $= 4/4 = 1$), $ss\ ps \to dn$ (confidence $= 4/4 = 1$), $dn\ ps \to ss$ (confidence $= 4/4 = 1$).

Even the generation of all the association rules with a given confidence from the set of frequent itemsets may become computationally expensive. For each itemset (of size k), we need to test $2^k - 2$ potentially confident rules, because we must consider each subset of the itemset as an antecedent (except for the empty and the full itemset), and each such test requires access to the support of the antecedent as well as the consequent of the rule. Thus, we need access to the support of all subsets of each frequent itemset. On the other hand, k is often small compared to $|M|$.

The complete set of association rules with *minsup* = 0.33 and *minconf* = 0.66 for the enlarged planets context is shown in Figure 5.8 (we indicate the rules extracted from each frequent itemset and denote by an asterisk the rules with 100% confidence).

5.3.1 Mining frequent concepts

So far, we have not discussed the use of concept lattices for mining association rules. In fact, frequent itemsets and *frequent* concepts (i.e., those concepts whose extent has a cardinality of at least $minsup \cdot |G|$) are closely related. The key observation is that the support of any itemset Y is equal to the support of its closure Y'', which guarantees that all frequent itemsets are uniquely determined by frequent concepts.

In practice, one frequent concept will typically encode the information associated with several frequent itemsets. If two itemsets W and Z describe exactly the same set of objects, i.e. $W' = Z'$, then they will have the same support, which is the support of the concept (W'', W') or (Z'', Z') to which they belong. In the data mining literature, the (intents of) concepts are usually called *closed* itemsets, as they are closed under the closure doubleprime operator.

It may thus be more convenient to mine frequent concepts rather than frequent itemsets. The set of frequent concepts is a subset of the set of frequent itemsets, and it is usually much smaller, at least for dense contexts. To illustrate, consider again the enlarged planets context; while the set of all frequent itemsets contains 15 elements, the set of frequent concepts, shown in Figure 5.9, contains only seven elements.

Before addressing the problem of generating confident rules from frequent concepts, we describe an algorithm, called *Frequent Next Neighbours*, for mining the set of frequent concepts associated with a given context along with the Hasse diagram. This algorithm can be seen as a natural extension of the methods presented in Chapter 2, in particular of the Next Neighbours algorithm described in Section 2.1.3, to the problem of determining only the frequent concepts rather than all concepts.

As with the Next Neighbours algorithm, we start at the top element of the lattice (G, G') and build one level at a time, where the next level contains the children of all concepts present in the current level. This time, however, the algorithm halts as soon as the next level contains no frequent concepts, because more specific concepts will also not be frequent.

The use of support allows us to avoid the generation not only of levels with no frequent concepts but also of non-frequent concepts during the computation of the frequent lower neighbours of any frequent concept.

Instead of first computing all the lower neighbours, with the *FindLowerNeighbours* function used in the Next Neighbours algorithm, and then

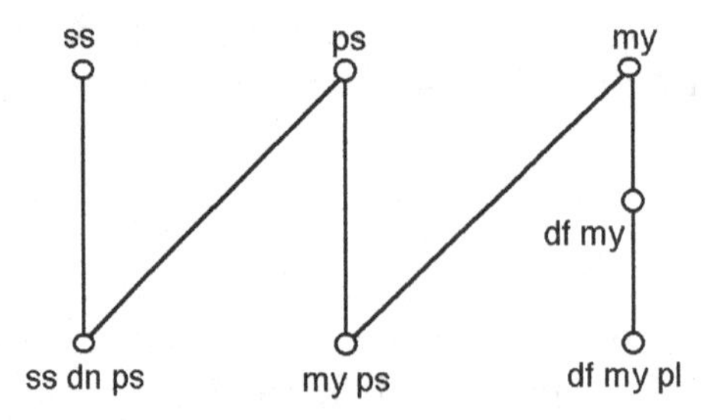

Figure 5.9 Hasse diagram of frequent concepts (*minsup* = 0.33) for the enlarged planets context.

```
FrequentNextNeighbours
Input: Context (G, M, I), minsup
Output: The join semilattice (C, E) of the frequent concepts of (G, M, I)

1. C := {(G, G')}
2. E := ∅;
3. currentLevel:= {(G, G')}
4. while currentLevel ≠ ∅
5.    nextLevel:= ∅
6.    for each (X, Y) ∈ currentLevel
7.       freqLowerNeighbours := FindFreqLowerNeighbours ((X, Y))
8.       for each (X1, Y1) ∈ freqLowerNeighbours
9.          if (X1, Y1) ∉ C then
10.            C := C ∪ {(X1, Y1)}
11.            nextLevel := nextLevel ∪{(X1, Y1)}
12.         Add edge (X1, Y1) → (X, Y) to E
13.   currentLevel:= nextLevel

function FindFreqLowerNeighbours((X, Y))
/* Returns the frequent lower neighbours of a concept */
1. candidates := ∅
2. maxGenCandidates := ∅
3. for each m ∈ M\Y
4.    sup := |{m}' ∩ X|
5.    if sup ≥ minsup then
6.       X1 := (Y ∪ {m})'
7.       Y1 := X1'
8.       if (X1, Y1) ∉ candidates
          then
9.          Add (X1, Y1) to candidates
10.         count(X1, Y1) := 1
          else
11.         count(X1, Y1) := count(X1, Y1) + 1
12.      if (|Y1| − |Y|) = count(X1, Y1) then
13.         Add (X1, Y1) to maxGenCandidates
14.return maxGenCandidates
```

Figure 5.10 Frequent Next Neighbours algorithm.

removing the non-frequent ones, we test whether each candidate neighbour has enough support before generating it. This is done simply by taking the intersection between the current concept's extent and each attribute extent and verifying that its cardinality is greater than *minsup*. The advantage is that in this way we do not need to compute the doubleprime operator for those candidate neighbours that would be discarded because not frequent.

A complete description of the algorithm is provided in Figure 5.10. To select the maximally general elements among the candidate frequent lower neighbours we use the same method as described for the Concepts Cover (Section 2.1.3) and Nearest Neighbours (Section 2.1.4) algorithms.

An auxiliary search tree can be used to check whether a concept has already been generated (line 9 of the main function in Figure 5.10), similar to the efficient implementation of the Next Neighbours algorithm. In this case, the worst-case time complexity of the Frequent Next Neighbours algorithm is given by the product of the number of invocations of the function *FindFreqLowerNeighbours* (which is equal to the total number of frequent concepts F) and the cost of computing the frequent lower neighbours. The most demanding operation of the latter is the calculation of the prime operator in lines 3 and 4 of the function *FindFreqLowerNeighbours*, which takes $O(|G||M|)$ time, performed at most $|M|$ times. The overall complexity is thus $O(F|G||M|^2)$.

In practice, however, the algorithm may be considerably more efficient, because the prime operator within the function *FindFreqLowerNeighbours* will typically be skipped for most of the attributes due to the preceding support test. The computation of support via set intersection (line 4 of function *FindFreqLowerNeighbours*) takes only $O(|G|)$ time if we use a bit-vector representation for each attribute's extent and for the extent of the concept for which the function *FindFreqLowerNeighbours* has been invoked.

Let us now see the algorithm at work on the enlarged planets context. The computation of the complete set of frequent concepts along with the line diagram is depicted in Figure 5.11. The working of the inner function is illustrated in Figure 5.12 for the concept '(Me V E Ma J S, *ps*)'.

Note that the algorithm constructs a join semilattice of the frequent concepts associated with the context. In this case, the top element of the semilattice has an empty intent, and thus does not contribute to the generation of the set of association rules. If we omit the top element, we obtain exactly the Hasse diagram shown in Figure 5.9.

5.3.2 Generating confident rules from frequent concepts

Once we have constructed the Hasse diagram of the frequent concepts, we are left with the problem of extracting the confident rules from it. In the following, we will consider the determination of rules with confidence equal to 100% separately from the determination of rules with confidence less than 100%.

The first task corresponds to finding the set of *implications* that can be extracted from the set of frequent concepts, and we have already studied this problem in Section 5.1. It is sufficient to run the Find Implications algorithm on the Hasse diagram of the frequent concepts.

For the enlarged planets context, the application of the Find Implications algorithm to the frequent concepts in Figure 5.9 returns the following reduced set of the frequent association rules with 100% confidence:

$$\{dn \rightarrow ss\ ps,\ ss\ ps \rightarrow dn,\ pl \rightarrow df\ my,\ df \rightarrow my\}.$$

C = {(Me V E Ma J S U N P, ∅)}
currentLevel = {(Me V E Ma J S U N P, ∅)}
nextLevel = ∅
freqLowerNeighbours((Me V E Ma J S U N P, ∅)) = {(Me V E Ma P, *ss*),
(Me V E Ma J S, *ps*), (E Ma J S U N P, *my*)}
C = {(Me V E Ma J S U N P, ∅), (Me V E Ma P, *ss*)}
nextLevel = {(Me V E Ma P, *ss*)}
Add edge (Me V E Ma P, *ss*) → (Me V E Ma J S U N P, ∅)
C = {(Me V E Ma J S U N P, ∅), (Me V E Ma P, *ss*), (Me V E Ma J S, *ps*)}
nextLevel = {(Me V E Ma P, *ss*), (Me V E Ma J S, *ps*)}
Add edge (Me V E Ma J S, *ps*) → (Me V E Ma J S U N P, ∅)
C = {(Me V E Ma J S U N P, ∅), (Me V E Ma P, *ss*), (Me V E Ma J S, *ps*),
(E Ma J S U N P, *my*)}
nextLevel = {(Me V E Ma P, *ss*), (Me V E Ma J S, *ps*), (E Ma J S U N P, *my*)}
Add edge (E Ma J S U N P, *my*) → (Me V E Ma J S U N P, ∅)
currentLevel = {(Me V E Ma P, *ss*), (Me V E Ma J S, *ps*), (E Ma J S U N P, *my*)}
nextLevel = ∅
freqLowerNeighbours((Me V E Ma P, *ss*)) = {(Me V E Ma, *ss dn ps*)}
C = {(Me V E Ma J S U N P, ∅), (Me V E Ma P, *ss*), (Me V E Ma J S, *ps*),
(E Ma J S U N P, *my*), (Me V E Ma, *ss dn ps*) }
nextLevel = {(Me V E Ma, *ss dn ps*)}
Add edge (Me V E Ma, *ss dn ps*) → (Me V E Ma P, *ss*)
freqLowerNeighbours((Me V E Ma J S, *ps*)) = {(Me V E Ma, *ss dn ps*), (E Ma J S, *my ps*)}
Add edge (Me V E Ma, *ss dn ps*) → (Me V E Ma J S, *ps*)
C = {(Me V E Ma J S U N P, ∅), (Me V E Ma P, *ss*), (Me V E Ma J S, *ps*),
(E Ma J S U N P, *my*), (Me V E Ma, *ss dn ps*), (E Ma J S, *my ps*) }
nextLevel = {(Me V E Ma, *ss dn ps*), (E Ma J S, *my ps*)}
freqLowerNeighbours((E Ma J S U N P, *my*)) = {(E Ma J S, *my ps*), (J S U N P, *df my*)}
Add edge (E Ma J S, *my ps*) → (E Ma J S U N P, *my*)
C = {(Me V E Ma J S U N P, ∅), (Me V E Ma P, *ss*), (Me V E Ma J S, *ps*),
(E Ma J S U N P, *my*), (Me V E Ma, *ss dn ps*), (E Ma J S, *my ps*),
(J S U N P, *df my*)}
nextLevel = {(Me V E Ma, *ss dn ps*), (E Ma J S, *my ps*), (J S U N P, *df my*)}
Add edge (J S U N P, *df my*) → (E Ma J S U N P, *my*)
currentLevel = {(Me V E Ma, *ss dn ps*), (E Ma J S, *my ps*), (J S U N P, *df my*)}
nextLevel = ∅
freqLowerNeighbours((Me V E Ma, *ss dn ps*)) = ∅
freqLowerNeighbours((E Ma J S, *my ps*)) = ∅
freqLowerNeighbours((J S U N P, *df my*)) = {(U N P, *df my pl*)}
C = {(Me V E Ma J S U N P, ∅), (Me V E Ma P, *ss*), (Me V E Ma J S, *ps*),
(E Ma J S U N P, *my*), (Me V E Ma, *ss dn ps*), (E Ma J S, *my ps*),
(J S U N P, *df my*), (U N P, *df my pl*)}
nextLevel = {(U N P, *df my pl*)}
Add edge (U N P, *df my pl*) → (J S U N P, *df my*)
currentLevel = {(U N P, *df my pl*)}
nextLevel = ∅
freqLowerNeighbours((U N P, *df my pl*)) = ∅
currentLevel = ∅

Figure 5.11 Step-by-step computation of the join semilattice of the frequent concepts (*minsup* = 0.33) for the context in Table 5.1 using the Frequent Next Neighbours algorithm.

FindFreqLowerNeighbours((Me V E Ma J S, *ps*))
ss: *sup* = | {Me V E Ma P} ∩ {Me V E Ma J S} | = 4
X_1 = {Me V E Ma}
Y_1 = {*ss dn ps*}
candidates = {(Me V E Ma, *ss dn ps*)}
count (Me V E Ma, *ss dn ps*) = 1
sm: *sup* = | {U N} ∩ {Me V E Ma J S} | = 0
sl: *sup* = | {J S} ∩ {Me V E Ma J S} | = 2
dn: *sup* = | {Me V E Ma} ∩ {Me V E Ma J S} | = 4
X_1 = {Me V E Ma}
Y_1 = {*ss dn ps*}
count (Me V E Ma, *ss dn ps*) = 2
maxGenCandidates = {(Me V E Ma, *ss dn ps*)}
df: *sup* = | {J S U N P} ∩ {Me V E Ma J S} | = 2
my: *sup* = | {E Ma J S U N P} ∩ {Me V E Ma J S} | = 4
X_1 = {E Ma J S}
Y_1 = {*myps*}
candidates = {(Me V E Ma, *ss dn ps*), (E Ma J S, *my ps*)}
count (E Ma J S, *my ps*) = 1
maxGenCandidates = {(Me V E Ma, *ss dn ps*), (E Ma J S, *my ps*)}
mn: *sup* = | {Me V} ∩ {Me V E Ma J S} | = 2
output = {(Me V E Ma, *ss dn ps*), (E Ma J S, *my ps*)}

Figure 5.12 Step-by-step computation of the frequent lower neighbours (*minsup* = 0.33) of the concept '(Me V E Ma J S, *ps*)'.

Note that the other eight frequent rules with 100% confidence shown in Figure 5.8 are redundant, because they can be obtained from some of the four rules shown above by shifting one attribute from the consequent to the antecedent of the rule.

Rules with confidence less than 100% can be read off from the Hasse diagram of the frequent concepts, but this time we need to look at the lower neighbours of each concept, instead of its upper neighbours as with 100% confidence rules.

For each edge connecting an attribute concept $\mu(m)$ to a lower neighbour of $\mu(m)$, a set of rules can be generated. The antecedents will be formed of all subsets of $\mu(m)$'s intent that contain m; the consequents will consist of all subsets of the intent of $\mu(m)$'s lower neighbour that contain some attribute not belonging to $\mu(m)$'s intent. All the rules derived from one edge will have the same confidence.

This process can be best illustrated using a slightly modified version of the set of frequent concepts from which the rules are extracted. In Figure 5.13 we show the Hasse diagram of the frequent concepts of the enlarged planets context using minimal labelling and adding its confidence to each edge. The confidence of an edge between two concepts is the ratio of the cardinality of the more specific concept's extent to the cardinality of the more general concept's extent.

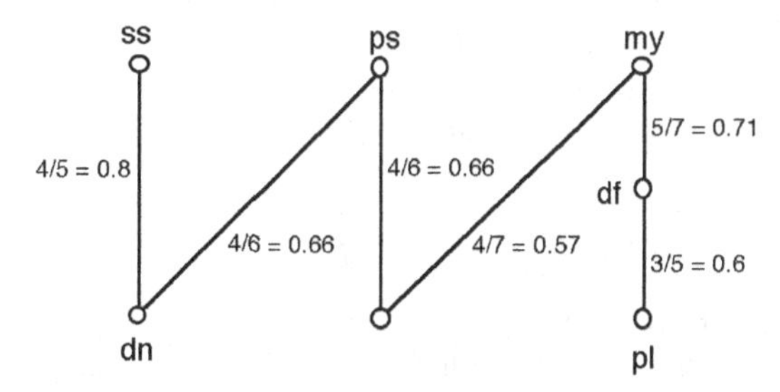

Figure 5.13 Hasse diagram of frequent concepts (*minsup* = 0.33) with minimal labelling and precision on the edges.

The edge between the concepts *my* and *df* in Figure 5.13 yields the rule $my \to df$, with a confidence of $5/7 = 0.71$. The edge between the concepts *df* and *pl* yields the rules $df \to pl$, $df\,my \to pl$, and $df \to pl\,my$, all with a confidence of $3/5 = 0.6$ (note that the full intents of concepts *df* and *pl* are (*df my*) and (*df my pl*), respectively).

Rules between non-adjacent concepts can be derived by transitivity, where the confidence of the derived rule is given by the product of the confidence of the generating rules. For example, from rules $my \to df$ (confidence $= 5/7$) and $df \to pl$ (confidence $= 3/5$) we obtain the rule $my \to df$, with a confidence of $5/7 \cdot 3/5 = 3/7$.

Using frequent concepts, it is not necessary to generate all the rules corresponding to each edge because some rules are redundant and can be derived from some other (non-redundant) rule. The rule $R \to S$ can be derived from the rule $P \to Q$ if $R'' = P''$ and $(R \cup S)'' = (P \cup Q)''$. For instance, of the three rules generated from the edge between concepts *ss* and *dn* (i.e., $ss \to ps$, $ss \to dn\,ps$, $ss \to dn$), the first two can be derived from the third because the condition previously stated is satisfied. The only non-redundant rule is thus $ss \to dn$ (note that the rule $dn \to ps$ holds with 100% confidence).

The rules with a given *minconf* can be generated by considering only the edges with confidence of at least *minconf*. The (reduced) set of association rules with confidence of at least 0.66 extracted from the frequent concepts is:

$$\{ss \to dn, ps \to dn, ps \to my, my \to df\}.$$

Note that in this way we generate four rules, whereas in Figure 5.8 there were eight rules with confidence less than 100%. In fact, the remaining rules are redundant and can be derived from the non-redundant ones, including the rules with 100% confidence.

To summarize, in Figure 5.14 we show the complete set of rules with confidence of at least 0.66 generated from the line diagram of the frequent concepts. Compared to the analogous set generated from the frequent itemsets, the number of rules falls from 20 to 8. The method illustrated here

$\{ss\ dn\ ps\}$:	$dn \to^* ss\ ps,\ ss\ ps \to^* dn$
$\{ss\}$:	$ss \to dn$
$\{ps\}$:	$ps \to my,\ ps \to dn$
$\{df\ my\ pl\}$:	$pl \to^* df\ my$
$\{df\ my\}$:	$df \to^* my$
$\{my\}$:	$my \to^* df$

Figure 5.14 The set of association rules with *minsup* = 0.33 and *minconf* = 0.66, extracted from the Hasse diagram of the frequent concepts.

produces a particular *basis* of the association rules that hold in a context; other approaches are referenced in Section 5.5.

The benefit of using concepts rather than itemsets for mining association rules is thus twofold: the efficiency of the rule generation process improves and the number of association rules decreases.

5.4 Classification rules

In this section we consider another classical probabilistic rule mining task, i.e., induction of classification rules. Intuitively, we seek expressions of the form 'any object with features a and b most likely belongs to class k'. This problem is usually cast as a (supervised) inductive learning task, in which the input is a set of objects (or instances) with class labels and the goal is to correctly classify new instances with a hidden class label.

In most inductive learning systems, the search through the space of rules to induce is guided by heuristics. Although these systems are usually effective, if the heuristics do not agree with the user's knowledge, the generated set of rules may not be easily understood. Furthermore, the use of heuristics may result in the omission of rules that are as plausible as those generated, or in a failure to include valuable rules that are relatively rare.

The concept lattice allows us to build a map through the space of rules, rather than trying to fit the training data heuristically. The concepts make the dependencies between attributes explicit, thus providing the basis for class prediction; however, as all such dependencies are theoretically justified they are also equally good, in a sense. This leads to the question which might be the best mechanism for extracting a minimum subset of classification rules from the concept lattice.

One approach is to derive for each class a set of *consistent* rules (i.e., which do not cover objects belonging to different classes) such that their union is *complete* (i.e., it covers all the objects belonging to that class). Such a derivation can be effectively performed using the concept lattice as a guide.

For a given context (G, M, I), let $c_1 \in M$ be the class attribute for which a description is sought. Recalling the inference properties of concept lattices introduced in Section 5.1, it is clear that all rules of the form $lhs \to c_1$

generated by the descendants of the attribute concept $\mu(c_1)$ are consistent. On the other hand, for each object belonging to c_1, it is possible to derive at least one consistent rule that covers the object from the descendants of $\mu(c_1)$ (in particular, from the object concept itself).

Hence, in order to find a consistent and complete description for class c_1, we can apply the Find Implications algorithm described in Section 5.1, with the two following restrictions. It is sufficient to consider only the concepts below $\mu(c_1)$ (or equal to $\mu(c_1)$), instead of the entire set of concepts (see line 1 in the main function in Figure 5.3), and to generate for each such concept only the rules having c_1 as consequent, thus instantiating the *lhsSet* variable to $Y \backslash c_1$ (see line 1 in the *FindImplicationsFromConcept*$((X, Y), L)$ function in Figure 5.3).

Consider, as an illustration, the problem of determining a description for the class 'distance far' (*df*) with respect to the objects in the modified planets context. Starting from the attribute concept of (*df*) in the concept lattice in Figure 5.2 (i.e., concept (J S P U N, *df my*)) and visiting all its descendants, application of the Find Implications algorithm returns the following implications: (*sl* → *df*) from the concept (J S, *sl df my ps*), (*pl* → *df*) from (P U N, *df my pl*), (*ss pl* → *df*) from (P, *ss df my pl*), (*sm* → *df*) from (U N, *sm df my pl*). Thus, a consistent and complete, although not minimal, description (in disjunctive normal form) of the 'distance far' class is:

sl OR *pl* OR *ss pl* OR *sm* → *df*.

Looking for a consistent and complete description is theoretically elegant, but it may not work in many situations. In particular, it is not suitable for noisy data, e.g., when an identical object has multiple class labels. In this case, there is no such thing as a consistent and complete description. We now describe a method that may account for noise and inconsistencies in the description of the data.

The first step consists of building the concept lattice of the set of training objects, ignoring the class attribute. Afterwards, the system is given access to the class label of the objects stored under each concept. For any test object, all concepts in the lattice are found that are more general than the object and mostly consistent. By mostly consistent we mean that the concepts may contain a small percentage of objects with a different class label, specified by a user-supplied parameter. Finally, the test object is assigned to the more numerous class in the set of concepts previously determined, following a majority vote decision.

This method can be efficiently implemented by a top-down breadth-first search through the relevant portion of the concept lattice. A detailed description of the algorithm is provided in Figure 5.15. For simplicity, we illustrate the case of binary classification, the extension to multiple classes being straightforward.

FindClass
Input: A concept lattice L of the context (G, M, I), a set of object classification functions $C(g)$ with $C(g) \in [c_1, c_2]$ for all $g \in G$, a noise parameter N, a new object x with intent $\{x\}'$
Output: Class of object x

1. $countC1 := 0$
2. $countC2 := 0$
3. $currentLevel := \{(G, G')\}$
4. **while** $currentLevel \neq \emptyset$
5. $\quad nextLevel := \emptyset$
6. $\quad$ **for** each $(X, Y) \in currentLevel$
7. $\quad\quad$ **if** $Y \subset \{x\}'$ **then**
8. $\quad\quad\quad$ **if** for all $g \in Y$, $C(g) = c_1$ (with tolerance N) **then**
9. $\quad\quad\quad\quad countC1 := countC1 + 1$
10. $\quad\quad\quad$ **if** for all $g \in Y$, $C(g) = c_2$ (with tolerance N) **then**
11. $\quad\quad\quad\quad countC2 := countC2 + 1$
12. $\quad\quad\quad$ **for** each unmarked (X_1, Y_1) child of (X, Y)
13. $\quad\quad\quad\quad$ Mark (X_1, Y_1)
14. $\quad\quad\quad\quad nextLevel := nextLevel \cup \{(X_1, Y_1)\}$
15. $\quad currentLevel := nextLevel$
16. **if** $countC1 > countC2$
17. $\quad$ **then return** c_1
17. $\quad$ **else return** c_2

Figure 5.15 Find Class algorithm.

Considering that each concept can be definitively assigned to one class (if any) through one single pass before examining the test set, and that for each test object only the concepts that are above the object's concept are examined (at most 2^k, where k is the number of attributes describing each object), the time required to classify p unseen object is given by $O(p2^k|M|)$. The last factor accounts for comparing the intents of a pair of concepts (line 7 in Figure 5.15).

5.5 Bibliographic notes

The theoretical results concerning the computational space complexity of implications were reported by Carpineto and Romano [45].

The algorithm for inferring a cover for implications from the concept lattice of the context is based on the work of Carpineto *et al.* [46]. An earlier, less efficient, algorithm than that presented here was described by Godin and Missaoui [103]; they also showed how to further exploit the information contained in the lattice to find smaller covers and suggested that it is possible to extend an incremental algorithm for concept lattice construction to generate both the concepts and the rules.

The problem of inferring a reduced (or most general) cover for implications has also received some attention. Two examples are the works by Ziarko [279] and Carpineto and Romano [45]. Ziarko's approach is based on a particular representation of the input context, called a *decision matrix*, from which the implications are extracted. Carpineto and Romano search a lattice-based rule space using 'positive' examples (i.e., the objects that contain the consequent of the rules to be learnt) and 'negative' examples (i.e., the objects that do not contain such a consequent). Such algorithms find more restricted covers than the concept lattice-based method presented here, usually at the cost of computational efficiency (especially considering the number of attributes).

The computation of a non-redundant cover for the implications that hold in a context is discussed in [41]. It is also useful to consider *embedded implications*, i.e., those implications which hold for subsets of attributes. Embedded implications can be efficiently extracted from a reduced cover of implications, or they can be inferred directly by adapting an algorithm for implication mining [234].

Implications are not only theoretically appealing, but also of practical interest. Obvious candidate applications are deterministic environments where we are seeking 'causal' associations, regardless of frequency, although a similar 'cause and effect' behaviour can also be observed in non-deterministic environments. The need to consider rules of extremely high confidence without regard for their support is advocated, for instance, by Cohen *et al.* [56] and Wang *et al.* [251]. They point out that such rules are a natural class of patterns in a number of applications, including copy detection, text mining and collaborative filtering. In such cases, it is likely that rules with low support and high confidence are interesting and provide new insights, whereas high-support rules will be obvious and well known.

Carpineto and Romano [45] observe that the size of the set of implications can remain substantially high even when mining classical test databases and considering only restricted covers; they argue that sometimes it may be more useful to concentrate on effective ways of pruning the set of implications (e.g., by focusing on short implications) rather than generating more powerful or approximate data dependencies.

The treatment of functional dependencies offered here is mostly based on on the work of Carpineto *et al.* [46]. The use of transformed contexts has been proposed by Wille [261]. The transformation of covers for implications into functional dependencies covers is analysed in depth by Missaoui and Godin [166].

Alternative methods for inferring functional dependencies have been presented in various fields, including rough sets ([178], [213]), artificial intelligence ([209], [20], [148]) and database theory ([66], [154], [155]). Interestingly, some of these use an intermediate representation based on pairs of objects whose attributes agree or disagree, in the manner of the information encoded in a transformed context.

Of central importance is the work of Mannila and Rähiä ([154], [155]), who proposed several algorithms for inferring functional dependencies, with the main goal of achieving a polynomial-time solution with the number of objects and the size of the smallest cover for the dependency set of the context. In general, however, while it is possible to operate in subquadratic time with the number of objects (or subexponential time with the number of attributes), it is difficult to find an algorithm that exhibits both behaviours. In this respect, the algorithm presented in this book is no exception, although its practical behaviour is quite good. A detailed experimental comparison between its performance and the performance of one of Mannila and Rähiä's algorithms is provided in [46].

Turning to association rules, the work of Agrawal and co-workers ([2], [4], [3]) has inspired much of the subsequent research in the field. The Apriori algorithm [3] and its many variants (e.g., [28], [180]) are based on the empirical assumption that itemsets of large size do not receive enough support from the data and therefore do not need be generated. However, as already mentioned, this behaviour holds for transactional data but not for dense, relational data, where a combinatorial explosion of large frequent itemsets has been observed.

The use of closed itemsets (or concepts) for mining association rules was first independently proposed by Zaki ([276], [273]), Pasquier *et al.* ([181], [182]) and Stumme [225]. Several algorithms for mining closed itemsets have been presented, including Close [181], Pascal [18], Closet [184], Titanic [229], Charm ([274], [275]), and Mafia [29].

The Frequent Next Neighbours algorithm presented in this book is novel. It differs from previous work on frequent concepts mining in that it directly exploits the structure of the join semilattice of the frequent concepts, whereas other algorithms are typically based on sophisticated data structures for representing the objects and the attributes ([184], [275]), or rely on specific pruning strategies for improving the Apriori-like generation of candidates ([18], [229]).

The set of frequent concepts can also be incrementally updated as new objects are added to the database; this issue is discussed in [186] and [242].

Another solution to the problem of mining large itemsets consists of generating only the maximal frequent itemsets using bidirectional search (e.g., [19], [29]). These are typically orders of magnitude fewer than all frequent itemsets, but this approach is not suitable for generating confident rules because subset frequency is not available.

The method used in this book to represent and extract the set of confident rules from frequent concepts is based on the work of Zaki ([276], [273]), who built upon Luxenburger's early work on *partial implications* [150]. Finding a useful basis of association rules is an open problem, because different theoretical frameworks are conceivable ([228], [226]). Furthermore, if we make particular assumptions about the description of the data, more powerful mechanisms for reducing the set of association rules without loss of information can be devised ([237], [179]).

The use of concept lattices to drive rule classification induction was one of the earliest data mining applications of concept data analysis. This was first suggested by Ganascia [89] and Oosthuizen and McGregor [177]. In a later paper [90], Ganascia also showed that other well-known top-down induction systems can be formalized using a concept lattice framework. More recently, Kuznetsov and Ganter investigated the relationships between minimal classification hypotheses and pseudo-intents [138].

The Find Class algorithm presented in this book is a refinement of an earlier one described by Carpineto and Romano in [34]. In the same paper, it is argued that it may be useful to use different methods for extracting classification rules from a concept lattice, depending on the properties exhibited by the data set at hand. Seeking maximally complete consistent concepts is one such additional method, which has also been proposed, with some variants, in [202].

As discussed above, rule classification mining based on concept lattices typically entails a search through an ordered hypothesis space, as opposed to using heuristic criteria to fit the training data. Examples of the former approach that do not rely on concept lattices include [194] and [252], whereas two well-known representatives of the latter approach are the AQ methodology developed by Michalski and co-workers (e.g., [164], [48]) and decision trees (e.g., [192]).

A different approach to concept lattice-based rule classification induction was proposed by Mephu Nguifo ([159], [160], [161]). Rather than generating a large concept lattice-based rule space from which to derive the classification rules sought, a subset of 'relevant' concepts are used to transform the original nominal attributes into a more restricted set of numeric attributes, with the goal of ruling out irrelevant features. This is a form of constructive induction driven by the concepts of the original objects.

A variety of techniques described in several recently published books (e.g., [116], [267], [163]) are available for performing rule classification mining. The approaches proposed span decision trees, neural networks, rough sets, Bayesian classifiers, cluster analysis, support vector machines, and boosting, to cite a few. For a focused discussion of the relative merits and drawbacks of *some* alternative rule mining approaches, including concept data analysis, see [267].

Concept lattices can also be combined with existing classification methods. Xie *et al.* describe an approach where a basic classifier such as Naive Bayes or Nearest Neighbour is trained on a pruned set of concepts built from the training objects rather than on the training objects themselves [269].

A task related to class prediction is *class discovery*, in which the goal is to discover useful partitions (or clusters) of a set of objects. This is also termed unsupervised learning, because no training set is provided. Assuming that the natural distribution of the objects to be partitioned is known, Carpineto and Romano [34] construct out of the set of concepts in the lattice all possible partitions satisfying the given distribution, and show that for a growing sample of input objects such partitions may converge to the 'natural' partition.

Besides automatic methods for rule mining, it is possible to acquire knowledge from a concept lattice in an interactive fashion [222]. A classical method is *attribute exploration* [91], best described in [96]. Assuming that the object set is only partly available, the idea is to iteratively provide the user with a set of current implications which can be either accepted or refuted (by stating counter-examples), until the concept lattice is completely determined. This approach can be extended to cope with situations where both the objects and the attributes are not completely known [223]. Even the generation of frequent concepts can benefit from an interactive approach, because the user may analyse the frequent concepts corresponding to a small subset of attributes and incrementally add new interesting frequent attributes [73].

References

[1] M. Agosti, M. Melucci, and F. Crestani. Automatic authoring and construction of hypertexts for information retrieval. *ACM Multimedia Systems*, 3:15–24, 1995.

[2] R. Agrawal, T. Imielinski, and A. Swami. Mining association rules between sets of items in large databases. In *ACM SIGMOD International Conference on Management of Data*, pp. 207–216, Washington, DC, USA, 1993.

[3] R. Agrawal, H. Mannila, R. Srikant, H. Toivonen, and A.I. Verkamo. Fast discovery of association rules. In U.M. Fayyad, G. Piatetsky-Shapiro, P. Smyth, and R. Uthurusamy, eds, *Advances in Knowledge Discovery and Data Mining*, pp. 307–328. AAAI Press, 1996.

[4] R. Agrawal and R. Srikant. Fast algorithms for mining association rules. In *Proceedings of the 20th Very Large Data Bases Conference*, pp. 487–499, Santiago, Chile, 1994.

[5] A.V. Aho, J.E. Hopcroft, and J.D. Ullman. *Data Structures and Algorithms*. Addison-Wesley, 1983.

[6] H. Alaoui. Algorithmes de manipulation du treillis de Galois d'une relation binaire et applications. Master's thesis, Université du Québec à Montréal, 1992.

[7] G. Amati, C. Carpineto, and G. Romano. FUB at TREC-10 Web track: A probabilistic framework for topic relevance term weighting. In *Proceedings of the 10th Text REtrieval Conference (TREC-10)*, NIST Special Publication 500-250, pp. 182–191, Gaithersburg, MD, USA, 2001.

[8] G. Amati, C. Carpineto, and G. Romano. Comparing weighting models for monolingual information retrieval. In *Working Notes for the CLEF 2003 Workshop*, pp. 169–178, Trondheim, Norway, 2003.

[9] G. Amati, C. Carpineto, and G. Romano. Fondazione Ugo Bordoni at TREC 2003: Robust and Web track. In *Proceedings of TREC 2003*, pp. 210–219, Gaithersburg, MD, USA, 2003.

Concept Data Analysis: Theory and Applications. Claudio Carpineto and Giovanni Romano
 ISBN: 0-470-85055-8

[10] G. Amati and C.J. van Rijsbergen. Probabilistic models of information retrieval based on measuring divergence from randomness. *ACM Transactions on Information Systems*, 20(4):357–389, 2002.

[11] K. Andrews, M. Pichler, and J. Wolte. Information pyramids: A new approach to visualizing large hierarchies. In *Proceedings of Late Breaking Hot Topics, IEEE Visualization '97*, pp. 49–52, Phoenix, AZ, USA, 1997.

[12] F. Baader. Computing a minimal representation of the subsumption lattice of all conjunctions of concept defined in a terminology. In *Proceedings of the International KRUSE Symposium*, pp. 168–178, Santa Cruz, CA, USA, 1995.

[13] R. Baeza-Yates and B. Ribeiro-Neto. *Modern Information Retrieval.* Addison-Wesley, 1999.

[14] L.D. Baker and A.K. McCallum. Distributional clustering of words for text classification. In *Proceedings of the 21st Annual International ACM SIGIR Conference on Research and Development in Information Retrieval*, pp. 96–103, Melbourne, Australia, 1998.

[15] M.Q.W. Baldonado, A. Woodruff, and A. Kuchinsky. Guidelines for using multiple views in information visualizations. In *Proceedings of AVI 2000*, pp. 110–119, Palermo, Italy, 2000.

[16] M. Barbut and B. Monjardet. *Ordre et Classification, Algèbre et Combinatoire*, 2 vols. Hachette, 1970.

[17] L. Bartram, A. Ho, and F. Henigman. The continuous zoom: A constrained fisheye technique for viewing and navigating large information spaces. In *Proceedings of ACM UIST'95 Conference*, pp. 207–215, Pittsburgh, PA, USA, 1995.

[18] Y. Bastide, R. Taouil, N. Pasquier, G. Stumme, and L. Lakhal. Mining frequent patterns with counting inference. *SIGKDD Explorations, Special Issue on Scalable Algorithms*, 2(2):71–80, 2000.

[19] R. Bayardo. Efficiently mining long patterns from databases. In *Proceedings of the 1998 ACM SIGMOD Conference on Management of Data*, pp. 85–93, Seattle, WA, USA, 1998.

[20] S. Bell and P. Brockhausen. Discovery of data dependencies in relational databases. In *Working Notes of the MLnet Workshop on Statistics, Machine Learning and Knowledge Discovery in Databases*, pp. 53–58, Heraklion, Crete, Greece, 1995.

[21] E. Berenci, C. Carpineto, V. Giannini, and S. Mizzaro. Effectiveness of keyword-based display and selection of retrieval results for interactive searches. *International Journal on Digital Libraries*, 3(3):249–260, 2000.

[22] A. Berry, J.-P. Bordat, and O. Cogis. Generating all the minimal separators of a graph. *International Journal of Foundations of Computer Science*, 11:397–404, 2000.

[23] A. Berry and A. Sigayret. Representing a concept lattice by a graph. In *Proceedings of the Workshop on Discrete Mathematics and Data Mining at the 2nd SIAM Conference on Data Mining*, Arlington, VA, USA, 2002.

[24] G. Birkhoff. *Lattice Theory*. American Mathematical Society, 1940.

[25] H. Blockeel, L. De Raedt, and J. Ramong. Top-down induction of clustering trees. In *Proceedings of the 15th International Conference on Machine Learning*, pp. 55–63, Madison, WI, USA, 1998.

[26] J.P. Bordat. Calcul pratique du treillis de Galois d'une correspondance. *Mathématiques et Sciences Humaines*, 96:31–47, 1986.

[27] G. Brajnik, S. Mizzaro, and C. Tasso. Evaluating user interfaces to information retrieval systems. In *Proceedings of the 19th Annual International ACM SIGIR Conference on Research and Development in Information Retrieval*, pp. 128–136, Zurich, Switzerland, 1996.

[28] S. Brin, R. Motwani, J.D. Ullman, and S. Tsur. Dynamic itemset counting and implication rules for market basket data. In *Proceedings of the ACM SIGKDD International Conference on Management of Data*, pp. 255–264, Tucson, AZ, USA, 1997.

[29] D. Burdick, M. Calimlim, and J. Gehrke. MAFIA: a maximal frequent itemset algorithm for transactional databases. In *Proceedings of the 17th International Conference on Data Engineering*, Heidelberg, Germany, 2001.

[30] D. Byrd. A scrollbar-based visualization for document navigation. In *Proceedings of the 4th ACM Digital Library Conference*, pp. 122–129, Berkeley, CA, USA, 1999.

[31] S. Card, T. Moran, and A. Newell. *The Psychology of Human–Computer Interaction*. Lawrence Erlbaum Associates, 1983.

[32] C. Carpineto. Trading off consistency and efficiency in version-space induction. In *Proceedings of the 9th International Conference on Machine Learning*, pp. 43–48, Aberdeen, UK, 1992.

[33] C. Carpineto, R. De Mori, G. Romano, and B. Bigi. An information theoretic approach to automatic query expansion. *ACM Transactions on Information Systems*, 19(1):1–27, 2001.

[34] C. Carpineto and G. Romano. An order-theoretic approach to conceptual clustering. In *Proceedings of the 10th International Conference on Machine Learning*, pp. 33–40, Amherst, MA, USA, 1993.

[35] C. Carpineto and G. Romano. Dynamically bounding browsable retrieval spaces: an application to Galois lattices. In *Proceedings of RIAO*

94: Intelligent Multimedia Information Retrieval Systems and Management, pp. 520–533, New York, USA, 1994.

[36] C. Carpineto and G. Romano. Galois-lattice clustering with a changing context. In *Working Notes of the AAAI-94 Workshop on Indexing and Reuse in Multimedia Systems*, pp. 164–174, Seattle, WA, USA, 1994.

[37] C. Carpineto and G. Romano. Automatic construction of navigable concept networks characterizing text databases. In M. Gori and G. Soda, eds, *Topics in Artificial Intelligence*, pp. 67–78. Springer, 1995.

[38] C. Carpineto and G. Romano. ULYSSES: A lattice-based multiple interaction strategy retrieval interface. In B. Blumenthal, J. Gornostaev, and C. Unger, eds, *Human Computer Interaction, 5th International Conference, EWHCI, Selected Papers*, pp. 91–104. Springer, 1995.

[39] C. Carpineto and G. Romano. Information retrieval through hybrid navigation of lattice representations. *International Journal of Human–Computer Studies*, 45(5):553–578, 1996.

[40] C. Carpineto and G. Romano. A lattice conceptual clustering system and its application to browsing retrieval. *Machine Learning*, 24(2):1–28, 1996.

[41] C. Carpineto and G. Romano. Inferring minimal rule covers from relations. In *Proceedings of the 5th Congress of the Italian Association for Artificial Intelligence*, pp. 147–158, Rome, Italy, 1997.

[42] C. Carpineto and G. Romano. Effective reformulation of Boolean queries with concept lattices. In *Proceedings of the 3rd International Conference on Flexible Query-Answering Systems*, pp. 83–94, Roskilde, Denmark, 1998.

[43] C. Carpineto and G. Romano. Towards better techniques for automatic query expansion. In *Proceedings of the 3rd European Conference on Digital Libraries (ECDLM-U99)*, pp. 126–141, Paris, France, 1999.

[44] C. Carpineto and G. Romano. Order-theoretical ranking. *Journal of the American Society for Information Science*, 51(7):587–601, 2000.

[45] C. Carpineto and G. Romano. Mining short-rule covers in relational databases. *Computational Intelligence*, 19(3):215–234, 2003.

[46] C. Carpineto, G. Romano, and P. d'Adamo. Inferring dependencies from relations: A conceptual clustering approach. *Computational Intelligence*, 15(4):415–441, 1999.

[47] C. Carpineto, G. Romano, and V. Giannini. Improving retrieval feedback with multiple term-ranking function combination. *ACM Transactions on Information Systems*, 20(3):259–290, 2002.

[48] G. Cervone, L.A. Panait, and R.S. Michalski. The development of the AQ20 learning system and initial experiments. In *Proceedings of the 10th International Symposium on Intelligent Information Systems*, pp. 13–29, Zakopane, Poland, 2001.

[49] S. Chakrabarti. *Mining the Web*. Morgan Kaufmann, 2003.

[50] M Chalmers and P. Chitson. BEAD: explorations in information visualization. In *Proceedings of the 15th Annual International ACM SIGIR Conference on Research and Development in Information Retrieval*, pp. 330–337, Copenhagen, Denmark, 1996.

[51] L. Chaudron and N. Maille. Generalized formal concept analysis. In *Proceedings of the 8th International Conference on Conceptual Structures*, pp. 357–370, Darmstadt, Germany, 2000.

[52] L. Chaudron, N. Maille, and M. Boyer. The cube lattice model and its applications. *Applied Artificial Intelligence*, 17(3):207–242, 2003.

[53] M. Chein. Algorithme de recherche des sous-matrices premières d'une matrice. *Bulletin Mathématique de la Société des Sciences Mathématiques de la République Socialiste de Roumanie*, 13:21–25, 1969.

[54] J.M. Cigarrän, J. Gonzalo, A. Peñas, and F. Verdejo. Browsing search results via formal concept analysis: selection of attributes and concept descriptors. In *Proceedings of the 2nd International Conference on Formal Concept Analysis (ICFCA04)*, Sydney, Australia, 2004.

[55] C. Cleverdon. Optimizing convenient on-line access to bibliographic databases. *Information Service and Use*, 4(1):37–47, 1974.

[56] E. Cohen, M. Datar, S. Fujiwara, A. Gionis, P. Indyk, R.J. Motwani, and J.D. Ullman. Finding interesting associations without support pruning. In *Proceedings of the 16th International Conference on Data Engineering*, pp. 489–499, San Diego, CA, 2000.

[57] R. Cole. Automated layout of concept lattices using force directed placement and genetic algorithms. In *Proceedings of the 23rd Australasian Computer Science Conference*, pp. 47–53, Canberra, Australia, 2000.

[58] R. Cole. The management and visualization of document collections using formal concept analysis. PhD thesis, Griffith University, Brisbane, Australia, 2000.

[59] R. Cole and P. Eklund. Text retrieval for medical discharge summaries using SNOMED and formal concept analysis. In *Proceedings of the First Australian Document Computing Symposium (ADCS)*, pp. 50–58, Melbourne, Australia, 1996.

[60] R. Cole and P. Eklund. Scalability in formal concept analysis. *Computational Intelligence*, 15(1):11–27, 1999.

[61] R. Cole and P. Eklund. Browsing semi-structured web texts using formal concept analysis. In *Proceedings of the 9th International Conference on Conceptual Structures*, pp. 319–332, Stanford, CA, USA, 2001.

[62] R. Cole, P. Eklund, and G. Stumme. Document retrieval for e-mail search and discovery using formal concept analysis. *Applied Artificial Intelligence*, 17(3):257–280, 2003.

[63] R. Cole and G. Stumme. CEM: A conceptual email manager. In *Proceedings of the 8th International Conference on Conceptual Structures*, pp. 438–452, Darmstadt, Germany, 2000.

[64] W.R. Cook. Interfaces and specifications for the Smalltalk-80 collection classes. In *Proceedings of the 7th Annual Conference on Object Oriented Programming Systems Languages and Applications*, pp. 1–15, Vancouver, BC, Canada, 1992.

[65] J. Cooper and R. Byrd. Lexical navigation: visually prompted query expansion and refinement. In *Proceedings of the 2nd ACM Digital Library Conference*, pp. 237–246, Philadelphia, USA, 1997.

[66] S. Cosmakadis, P. Kanellakis, and N. Spyratos. Partition semantics for relations. *Journal of Computer and System Sciences*, 33:203–233, 1986.

[67] T.M. Cover and J.A. Thomas. *Elements of Information Theory*. Wiley, 1991.

[68] P. Crawley and R.P. Dilworth. *Algebraic Theory of Lattices*. Prentice Hall, 1973.

[69] B. Davey and H. Priestley. *Introduction to Lattices and Order*. Cambridge University Press, 1990.

[70] T.R. Davies and S.J. Russel. A logical approach to reasoning by analogy. In *Proceedings of the 10th International Joint Conference on Artificial Intelligence*, pp. 264–270, Milan, Italy, 1987.

[71] S. Deerwester, S.T. Dumais, W. Furnas, T.K. Landauer, and R. Harshman. Indexing by latent semantic analysis. *Journal of the American Society for Information Science*, 41(6):391–407, 1990.

[72] Y. Dhillon, S. Manella, and R. Kumar. Divisive information-theoretic feature clustering algorithm for text classification. *Journal of Machine Learning Research*, 3, 2003.

[73] L. Dumitriu. Interactive mining and knowledge reuse for the closed-itemset incremental mining problem. *SIGKDD Explorations*, 3(2):28–36, 2002.

[74] P. Eades and R. Tamassia. Algorithms for drawing graphs: an annotated bibliography. Technical Report CS-89-90, Department of Computer Science, Brown University, 1989.

[75] L. Egghe and R. Rousseau. Topological aspects of information retrieval. *Journal of the American Society for Information Science*, 49(13):1144–1160, 1998.

[76] B. S. Everitt. *Cluster Analysis* (third edition). Arnold, 1993.

[77] M. Faïd, R. Missaoui, and R. Godin. Knowledge discovery in complex objects. *Computational Intelligence*, 15(1):28–49, 1999.

[78] S. Ferré and D.R. King. BLID: an application of logical information systems to bioinfomatics. In *Proceedings of the 2nd International Conference on Formal Concept Analysis (ICFCA04)*, Sydney, Australia, 2004.
[79] S. Ferré and O. Ridoux. A file system based on concept analysis. In *Proceedings of the 1st International Conference on Computational Logic*, pp. 1033–1047, London, UK, 2000.
[80] S. Ferré and O. Ridoux. A logical generalization of formal concept analysis. In *Proceedings of the 8th International Conference on Conceptual Structures*, pp. 371–384, Darmstadt, Germany, 2000.
[81] D. Fisher. Knowledge acquisition via incremental conceptual clustering. *Machine Learning*, 2:139–172, 1987.
[82] D. H. Fisher. Data mining tasks and methods. Clustering: conceptual clustering. In W. Klosgen and J. M. Zytkow, eds, *Handbook of Data Mining and Knowledge discovery*, pp. 388–396. Oxford University Press, 2002.
[83] R. Freese, J. Jezek, and J.B. Nation. *Free Lattices*. Mathematical Surveys and Monographs, 42. American Mathematical Society, 1995.
[84] H. Fu and E. Mephu Nguifo. Partitioning large data to scale-up lattice-based algorithms. In *Proceedings of the 15th International Conference on Tools with Artificial Intelligence*, Washington, DC, USA, 2003.
[85] C.-H. Fung and Q. Li. Advanced conceptual clustering and associated querying facilities in object-oriented databases. *International Journal of Software Engineering and Knowledge Engineering*, 9(3):343–368, 1999.
[86] P. Funk, A. Lewien, and G. Snelting. Algorithms for concept lattice decomposition and their applications. Technical Report 95-05, Fachbereich Informatik, TU Braunschweig, Germany, 1998.
[87] G.W. Furnas. Experience with an adaptive indexing scheme. In *Proceedings of ACM CHI'85 Conference on Human Factors in Computing Systems*, pp. 130–135, San Francisco, USA, 1985.
[88] G.W. Furnas. Generalized fisheye views. In *Proceedings of ACM CHI'86 Conference on Human Factors in Computing Systems*, pp. 16–23, Boston, USA, 1986.
[89] J.G. Ganascia. CHARADE: A rule system learning system. In *Proceedings of the 10th International Joint Conference on Artificial Intelligence*, pp. 345–347, Milan, Italy, 1987.
[90] J.G. Ganascia. TDIS: an algebraic formalization. In *Proceedings of the 13th International Joint Conference on Artificial Intelligence*, pp. 1008–1013, Chambéry, France, 1993.
[91] B. Ganter. Two basic algorithms in concept analysis. Technical Report FB4–Preprint No. 831, TU Darmstadt, Germany, 1984.

[92] B. Ganter. Composition and decomposition in formal concept analysis. In H.-H. Bock, ed., *Classification and Related Methods of Data Analysis*, pp. 561–566. North-Holland, 1988.

[93] B. Ganter. Dependencies of many-valued attributes. In H.-H. Bock, ed., *Classification and Related Methods of Data Analysis*, pp. 581–586. North-Holland, 1988.

[94] B. Ganter and S.O. Kuznetsov. Pattern structures and their projections. In *Proceedings of the 9th International Conference on Conceptual Structures*, pp. 129–142, Stanford, CA, USA, 2001.

[95] B. Ganter and R. Wille. Conceptual scaling. In F.S. Roberts, ed., *Applications of Combinatorics and Graph Theory to the Biological and Social Sciences*, pp. 139–167. Springer, 1989.

[96] B. Ganter and R. Wille. *Formal Concept Analysis–Mathematical Foundations*. Springer, 1999.

[97] J. Gennari, P. Langley, and D. Fisher. Models of incremental concept formation. *Artificial Intelligence*, 40:12–61, 1989.

[98] N. Gershon, S.K. Card, and S.G. Eick. Information visualization tutorial. In *Proceedings of ACM CHI'98: Human Factors in Computing Systems*, pp. 109–110, Los Angeles, USA, 1998.

[99] D.K. Gifford, P. Jouvelot, M.A. Sheldon, and J.W. O'Toole Jr. Semantic file systems. In *Proceedings of the 13th ACM Symposium on Operating Systems Principles*, pp. 16–25, 1991.

[100] R. Godin, J. Gecsei, and C. Pichet. Design of a browsing interface for information retrieval. In *Proceedings of the 12th Annual International ACM SIGIR Conference on Research and Development in Information Retrieval*, pp. 32–39, Cambridge, MA, USA, 1989.

[101] R. Godin and H. Mili. Building and maintaining analysis level class hierarchies using Galois lattices. In *Proceedings of the 8th Annual Conference on Object Oriented Programming Systems Languages and Applications*, pp. 394–410, Washington, DC, USA, 1993.

[102] R. Godin, H. Mili, G.W. Mineau, R. Missaoui, A. Arfi, and T.-T. Chau. Design of class hierarchies based on concept (Galois) lattices. *Theory and Practice of Object Systems*, 4(2):117–134, 1998.

[103] R. Godin and R. Missaoui. An incremental concept formation approach for learning from databases. *Theoretical Computer Science*, 133:387–419, 1994.

[104] R. Godin, R. Missaoui, and H. Alaoui. Learning algorithms using a Galois lattice structure. In *Proceedings of the 3rd International Conference on Tools with Artificial Intelligence*, pp. 22–29, San Jose, CA, USA, 1991.

[105] R. Godin, R. Missaoui, and H. Alaoui. Incremental concept formation algorithms based on Galois lattices. *Computational Intelligence*, 11(2):246–267, 1995.

[106] R. Godin, R. Missaoui, and A. April. Experimental comparison of navigation in a Galois lattice with conventional information retrieval methods. *International Journal of Man–Machine Studies*, 38:747–767, 1993.

[107] R. Godin, E. Saunders, and J. Gecsei. Lattice model of browsable data spaces. *Journal of Information Sciences*, 40:89–116, 1986.

[108] G. Golovchinsky and N.J. Belkin. Innovation and evaluation of information–a CHI98 workshop. *SIGCHI Bulletin*, 31(1):23–25, 1999.

[109] M.C. Golumbic. *Algorithmic Graph Theory and Perfect Graphs*. Academic Press, 1980.

[110] B. Gopal and U. Manber. Integrating content-based access mechanisms with hierarchical file systems. In *Proceedings of 3rd Symposium on Operating Systems Design and Implementation*, pp. 265–278, New Orleans, LA, USA, 1999.

[111] M. Graham. Visualising multiple overlapping classification hierarchies. PhD thesis, Napier University, Edinburgh, UK, 2001.

[112] M. Graham and J. Kennedy. Combining linking & focusing techniques for a multiple hierarchy visualisation. In *Proceedings of Information Visualization 2001*, pp. 425–432, London, UK, 2001.

[113] B. Groh, S. Strahringer, and R. Wille. TOSCANA-systems based on thesauri. In *Proceedings of the 6th International Conference on Conceptual Structures*, pp. 127–138, Montpellier, France, 1998.

[114] A. Guénoche. Construction du treillis de Galois d'une relation binaire. *Mathématiques, Informatique et Sciences Humaines*, 109:41–53, 1990.

[115] J.-L. Guiges and V. Duquenne. Familles minimales d'implications informatives resultant d'un tableau de données binaires. *Mathématiques et Sciences Humaines*, 95:5–18, 1986.

[116] J. Han and M. Kamber. *Data Mining: Concepts and Techniques*. Morgan Kaufmann, 2000.

[117] S. Hanson and M. Bauer. Conceptual clustering, categorization, and polymorphy. *Machine Learning*, 3:343–372, 1989.

[118] D. Harel and Y. Koren. A fast multi-scale method for drawing large graphs. In *Proceedings of Graph Drawing 2000*, pp. 183–196, Williamsburg, VA, USA, 2000.

[119] D. Harman. Relevance feedback and other query modification techniques. In W.B. Frakes and R. Baeza-Yates, eds, *Information Retrieval–Data Structures and Algorithms*, pp. 241–263. Prentice Hall, 1992.

[120] D.J. Harper and C.J. van Rijsbergen. An evaluation of feedback in document retrieval using co-occurrence data. *Journal of Documentation*, 34(3):189–216, 1978.

[121] M. Hearst. TileBars: Visualization of term distribution information in full text information access. In *Proceedings of ACM CHI'95: Human Factors in Computing Systems*, pp. 59–66, Denver, CO, USA, 1995.

[122] M. Hearst. User interfaces and visualization. In R. Baeza-Yates and B. Ribeiro-Neto, eds, *Modern Information Retrieval*, pp. 257–322. ACM Press, 1999.

[123] M.A. Hearst and J.O. Pedersen. Reexamining the cluster hypothesis: Scatter/Gather on retrieval results. In *Proceedings of the 19th Annual International ACM SIGIR Conference on Research and Development in Information Retrieval*, pp. 76–84, Zurich, Switzerland, 1996.

[124] M. Hemmje, C. Kunkel, and A. Willet. LyberWorld–A visualization user interface supporting full text retrieval. In *Proceedings of the 17th Annual International ACM SIGIR Conference on Research and Development in Information Retrieval*, pp. 249–259, Dublin, Ireland, 1996.

[125] I. Herman, G. Melançon, and M.S. Marshall. Graph visualization and navigation in information visualization: a survey. *IEEE Transactions on Visualization and Computer Graphics*, 6(1):24–43, 2000.

[126] D. Hiemstra and W. Kraaij. Twenty-one at TREC-7: Ad hoc and cross-language track. In *Proceedings of the 7th Text REtrieval Conference (TREC-7)*, NIST Special Publication 500-242, pp. 227–238, Gaithersburg, MD, USA, 1998.

[127] A. Hotho and G. Stumme. Conceptual clustering of text clusters. In *Proc. Fachgruppentreffen Maschinelles Lernen (FGML 2002)*, pp. 37–45, Hanover, Germany, 2002.

[128] A.K. Jain, M.N. Murty, and P.J. Flynn. Data clustering: A review. *ACM Computing Surveys*, 31(3):264–323, 1999.

[129] B. Johnson and B. Shneiderman. Treemaps: A space-filling approach to the visualization of hierarchical information structures. In *Proceedings of IEEE Visualization '91*, pp. 284–291, San Diego, CA, USA, 1991.

[130] D. Karp, Y. Schabes, M. Zaidel, and D. Egedi. A freely available wide coverage morphological analyzer for English. In *Proceedings of the 14th International Conference on Computational Linguistics (COLING'92)*, pp. 950–955, Nantes, France, 1992.

[131] G. Karpys and E.-H. Han. Fast supervised dimensionality reduction algorithm with applications to document categorization and retrieval. In *Proceedings of the 9th International Conference on Information and Knowledge Management*, pp. 12–19, McLean, VA, USA, 2000.

[132] T.A. Keahey and E.L. Robertson. Nonlinear magnification fields. In *Proceedings of IEEE InfoVis'97*, pp. 38–45, Phoenix, AZ, USA, 1997.

[133] T. Kloks and D. Kratsch. Listing all minimal separators of a graph. *SIAM Journal on Computing*, 27:605–613, 1998.

[134] D.E. Knuth. *The Art of Computer Programming, Volume 3, Sorting and Searching* (second edition). Addison-Wesley, 1998.

[135] D. Koller and M. Sahami. Toward optimal feature selection. In *Proceedings of the 13th International Conference on Machine Learning*, pp. 284–292, Bari, Italy, 1996.

[136] S.O. Kuznetsov. A fast algorithms for computing all intersections of objects in a finite semi-lattice. *Automatic Documentation and Mathematical Linguistics*, 27(5):11–21, 1993.

[137] S.O. Kuznetsov. On computing the size of a lattice and related decision problems in formal concept analysis. *Order*, 18(4):313–321, 2001.

[138] S.O. Kuznetsov and B. Ganter. Formalizing hypotheses with concepts. In *Proceedings of the 8th International Conference on Conceptual Structures*, pp. 340–354, Darmstadt, Germany, 2000.

[139] S.O. Kuznetsov and S.A. Obiedkov. Comparing performance of algorithms for generating concept lattices. *Journal of Experimental and Theoretical Artificial Intelligence*, 14(2–3):189–216, 2002.

[140] K.L. Kwok, L. Grunfeld, N. Dinstl, and P. Deng. TREC2003 Robust, Hard and QA track experiments using PIRCS. In *Proceedings of TREC 2003*, pp. 201–209, Gaithersburg, MD, USA, 2003.

[141] K.-S. Lee, Y.-C. Park, and K.-S. Choi. Re-ranking model based on document clusters. *Information Processing & Management*, 37(1):1–14, 2001.

[142] F. Lehmann and R. Wille. A triadic approach to formal concept analysis. In *Proceedings of the 5th International Conference on Conceptual Structures*, pp. 32–43, Santa Cruz, CA, USA, 1995.

[143] J. Lin. Divergence measures based on Shannon entropy. *IEEE Transactions on Information Theory*, 37(1):145–151, 1991.

[144] C. Lindig. Concept-based component retrieval. In *Working Notes of the IJCAI-95 Workshop: Formal Approaches to the Reuse of Plans, Proofs, and Programs*, pp. 21–25, Montreal, Canada, 1995.

[145] C. Lindig. Fast concept analysis. In *Working with Conceptual Structures–Contribution to the 8th International Conference on Conceptual Structures*, pp. 152–161, Darmstadt, Germany, 2000.

[146] M. Liquière and J. Sallantin. Structural machine learning with Galois lattices and graphs. In *Proceedings of the 15th International Conference on Machine Learning*, pp. 305–313, Madison, WI, USA, 1998.

[147] H. Liu and H. Motoda. *Feature Selection for Knowledge Discovery and Data Mining*. Kluwer Academic Publishers, 1998.

[148] S. Lopes, J.-M. Petit, and L. Lakhal. Functional and approximate dependency mining: database and FCA points of view. *Journal of Experimental and Theoretical Artificial Intelligence*, 14(2–3):93–114, 2002.

[149] D. Lucarella, S. Parisotto, and A. Zanzi. MORE: Multimedia Object Retrieval Environment. In *Proceedings of ACM Hypertext'93*, pp. 39–50, Seattle, WA, USA, 1993.

[150] M. Luxenburger. Implications partielles dans un contexte. *Mathématiques, Informatique et Sciences Humaines*, 29(113):35–55, 1991.

[151] Y. Maarek, D. Berry, and G. Kaiser. An information retrieval approach for automatically constructing software libraries. *IEEE Transactions on Software Engineering*, 17(8):800–813, 1991.

[152] D. Maier. *The Theory of Relational Databases*. Computer Science Press, 1983.

[153] Y. Malgrange. Recherche des sous-matrices premières d'une matrice à coefficients binaires: applications à certains problèmes de graphes. In *Deuxième Congrès de l'AFCALTI*, pp. 231–242, Gauthier-Villars, 1962.

[154] H. Mannila and K. Rähiä. Design by example: an application of Armstrong relations. *Journal of Computer and System Sciences*, 33:126–141, 1986.

[155] H. Mannila and K. Rähiä. Algorithms for inferring functional dependencies from relations. *Data & Knowledge Engineering*, 12(1):83–99, 1994.

[156] J. Martin and D. Billman. Acquiring and combining overlapping concepts. *Machine Learning*, 16(1–2):121–155, 1994.

[157] G. Melançon and I. Herman. DAG drawing from an information visualization perspective. In *Proceedings of VisSym'00–Joint Eurographics and IEEE TCVG Symposium on Visualization*, pp. 3–12, Amsterdam, Netherlands, 2000.

[158] C. Mellish. The description identification problem. *Artificial Intelligence*, 52(2):151–168, 1991.

[159] E. Mephu Nguifo. Galois lattice: a framework for concept learning, design, evaluation and refinement. In *Proceedings of the 6th International Conference on Tools with Artificial Intelligence*, pp. 461–467, New Orleans, LA, USA, 1994.

[160] E. Mephu Nguifo and P. Njiwoua. Using lattice-based framework as a tool for feature extraction. In *Proceedings of the 11th European Conference on Machine Learning*, pp. 304–309, Chemnitz, Germany, 1998.

[161] E. Mephu Nguifo and P. Njiwoua. IGLUE: A lattice-based constructive induction system. *International Journal of Intelligent Data Analysis*, 5(1):73–91, 2001.

[162] R.S. Michalski. Knowledge acquisition through conceptual clustering: a theoretical framework and an algorithm for partitioning data into conjunctive concepts. *Policy Analysis and Information Systems*, 4(3):219–244, 1980.

[163] R.S. Michalski, I. Bratko, and M. Kubat. *Machine Learning and Data Mining: Methods and Applications*. John Wiley & Sons, 1998.

[164] R.S. Michalski, I. Mozetic, J. Hong, and N. Lavrac. The multi-purpose incremental learning system AQ15 and its testing applications to three medical domains. In *Proceedings of the 5th National Conference on Artificial Intelligence*, pp. 1041–1047, Philadelphia, USA, 1986.

[165] R.S. Michalski and R. Stepp. Learning from observation: conceptual clustering. In R.S. Michalski, J.G. Carbonell, and T.M. Mitchell, eds, *Machine Learning: an Artificial Intelligence Approach, Vol. 1*, pp. 331–364. Tioga Publishing, 1986.

[166] R. Missaoui and R. Godin. Search for concepts and dependencies in databases. In W. Ziarko, ed., *Rough Sets, Fuzzy Sets and Knowledge Discovery*, pp. 16–23. Springer, 1994.

[167] T. Mitchell. Generalization as search. *Artificial Intelligence*, 18:203–226, 1982.

[168] T. Munzner. Exploring large graphs in 3D hyperbolic space. *IEEE Computer Graphics & Applications*, 18(4):18–23, 1998.

[169] D. Norman. Cognitive engineering. In D. Norman and S. Draper, eds, *User Centered System Design*, pp. 31–61. Lawrence Erlbaum Associates, 1986.

[170] C. North and B. Shneiderman. A taxonomy of multiple window co-ordinations. Technical Report CS-TR-3854, University of Maryland, 1997.

[171] C.L. North. A user-interface for co-ordinating visualization based on relational schemata: snap-together visualization. PhD thesis, Department of Computer Science, University of Maryland, College Park, MD, USA, 2000.

[172] L. Nourine and O. Raynaud. A fast algorithm for building lattices. *Information Processing Letters*, 71:199–204, 1999.

[173] L. Nourine and O. Raynaud. An on-line algorithm for building lattices. Technical Report 00041, LIRMM, Montpellier, France, 2000.

[174] G. D. Oosthuizen. The use of a lattice in knowledge processing. PhD thesis, University of Strathclyde, Glasgow, UK, 1988.

[175] G. D. Oosthuizen. Lattice-based knowledge discovery. In *Proceedings of AAAI-91 Knowledge Discovery in Databases Workshop*, pp. 221–235, Anaheim, CA, 1991.

[176] G. D. Oosthuizen. The application of concept lattices to machine learning. Technical Report CSTR 94/01, University of Pretoria, 1994.

[177] G. D. Oosthuizen and D.R. McGregor. Induction through knowledge base normalisation. In *Proceedings of the 8th European Conference on Artificial Intelligence*, pp. 396–401, Munich, Germany, 1988.

[178] M. Orlowska and M. Orlowski. Maintenance of knowledge in dynamic information systems. In R. Slowinski, ed., *Handbook of Applications and Advances of the Rough Sets Theory*, pp. 315–329. Kluwer Academic Publishers, 1992.

[179] B. Padmanabhan and A. Tuzhilin. Small is beautiful: discovering the minimal set of unexpected patterns. In *Proceedings of the ACM SIGKDD International Conference on Knowledge Discovery and Data Mining*, pp. 54–64, Boston, USA, 2000.

[180] J. Park, M. Chen, and P. Yu. Using a hash-based method with transaction trimming for mining association rules. *IEEE Transactions on Knowledge and Data Engineering*, 9(5):813–825, 1997.

[181] N. Pasquier, Y. Bastide, R. Taouil, and L. Lakhal. Discovering frequent closed itemsets for association rules. In *Proceedings of the 7th International Conference on Database Theory*, pp. 398–416, Jerusalem, Israel, 1999.

[182] N. Pasquier, Y. Bastide, R. Taouil, and L. Lakhal. Efficient mining of association rules using closed itemset lattices. *Journal of Information Systems*, 24(1):25–46, 1999.

[183] G. Pedersen. A browser for bibliographic information retrieval based on an application of lattice theory. In *Proceedings of the 16th Annual International ACM SIGIR Conference on Research and Development in Information Retrieval*, pp. 270–279, Pittsburgh, PA, USA, 1993.

[184] J. Pei, J. Han, and R. Mao. Closet: An efficient algorithm for mining frequent closed itemsets. In *Proceedings of the ACM SIGMOD International Workshop on Research Issues in Data Mining and Knowledge Discovery*, pp. 21–30, Dallas, TX, USA, 2000.

[185] N. Pernelle, M.-C. Rousset, H. Soldano, and V. Ventos. Zoom: a nested Galois lattice-based system for conceptual clustering. *Journal of Experimental and Theoretical Artificial Intelligence*, 14(2–3):157–187, 2002.

[186] J.L. Pfaltz and C.M. Taylor. Concept lattices as a scientific knowledge discovery technique. In *Proceedings of the 2nd SIAM International Conference on Data Mining*, pp. 65–74, Arlington, VA, USA, 2002.

[187] J. Ponte and W.B. Croft. A language modeling approach to information retrieval. In *Proceedings of the 21st Annual International ACM SIGIR Conference on Research and Development in Information Retrieval*, pp. 275–281, Melbourne, Australia, 1998.

[188] M.F. Porter. An algorithm for suffix stripping. *Program*, 14:130–137, 1980.

[189] J. Preece, Y. Rogers, H. Sharp, D. Benyon, S. Holland, and T. Carey. *Human–Computer Interaction*. Addison-Wesley, 1994.

[190] U. Priss. A graphical interface for document retrieval based on formal concept analysis. In *Proceedings of the 8th Midwest Artificial Intelligence and Cognitive Science Conference*, pp. 66–70, Dayton, OH, USA, 1997.

[191] U. Priss. Lattice-based information retrieval. *Knowledge Organization*, 27(3):132–142, 2000.

[192] R. Quinlan. *C4.5: Programs for Machine Learning*. Morgan Kaufmann, 1993.

[193] T.B. Rajashekar and B.W. Croft. Combining automatic and manual index representations in probabilistic retrieval. *Journal of the American Society for Information Science*, 46(4):272–283, 1995.

[194] P. Riddle, R. Segal, and O. Etzioni. Representation design and brute-force induction in a Boeing manufacturing domain. *Applied Artificial Intelligence*, 8(1):125–147, 1994.

[195] J.C. Roberts. On encouraging multiple views for visualization. In *Proceedings of IEEE InfoVis'98*, pp. 8–14, London, UK, 1998.

[196] G.G. Robertson, J.D. Mackinlay, and S.K. Card. Cone trees: Animated 3D visualizations of hierarchical information. In *Proceedings of CHI'91: Human Factors in Computing Systems*, pp. 189–194, New Orleans, LA, USA, 1991.

[197] S.E. Robertson. On term selection for query expansion. *Journal of Documentation*, 46(4):359–364, 1990.

[198] S.E. Robertson and K. Sparck Jones. Relevance weighting of search terms. *Journal of the American Society for Information Science*, 27:129–146, 1976.

[199] S.E. Robertson and S. Walker. Some simple effective approximations to the 2-Poisson method for probabilistic weighted retrieval. In *Proceedings of the 17th Annual International ACM SIGIR Conference on Research and Development in Information Retrieval*, pp. 311–317, Dublin, Ireland, 1994.

[200] S.E. Robertson, S. Walker, and M.M. Beaulieu. Okapi at TREC-7: Automatic Ad Hoc, Filtering, VLC, and Interactive track. In *Proceedings of the 7th Text REtrieval Conference (TREC-7)*, NIST Special Publication 500-242, pp. 253–264, Gaithersburg, MD, USA, 1998.

[201] C. Robinson. *Experiment, Design, and Statistics in Psychology*. Penguin, 1983.

[202] M. Sahami. Learning classification rules using lattices. In *Proceedings of the 8th European Conference on Machine Learning*, pp. 343–346, Heraklion, Crete, Greece, 1995.

[203] G. Salton. *Automatic Text Processing: The Transformation, Analysis, and Retrieval of Information by Computer*. Addison-Wesley, 1989.

[204] G. Salton and M. McGill. *Introduction to Modern Information Retrieval*. McGraw-Hill, 1983.

[205] M. Sarkar and M. Brown. Graphical fisheye views. *Communications of the ACM*, 37(12):73–84, 1994.

[206] I. Savnik and P. Flach. Bottom-up induction of functional dependencies from relations. In *Proceedings of the AAAI-93 Workshop on Knowledge Discovery in Databases*, pp. 174–185, Washington, DC, 1993.

[207] J. Savoy. Reports on CLEF-2001 experiments. In *Working Notes of CLEF 2001*, Darmstadt, Germany, 2001.

[208] J.C. Schlimmer. Learning determinations and checking databases. In *Proceedings of the AAAI-91 Workshop on Knowledge Discovery in Databases*, pp. 64–76, Anaheim, CA, USA, 1991.

[209] J.C. Schlimmer. Efficiently inducing determinations: A complete and systematic search algorithm that uses optimal pruning. In *Proceedings of the 10th International Conference on Machine Learning*, pp. 284–290, Amherst, MA, USA, 1993.

[210] D. Schütt. Abschätzungen für die Anzahl der Begriffe von Kontexten. Master's thesis, TH Darmstadt, Germany, 1998.

[211] C. Sherman and G. Price. *The Invisible Web: Uncovering Information Sources Search Engines Can't See*. CyberAge Books, 2001.

[212] B. Shneiderman. *Designing the User Interface: Strategies for Effective Human–Computer Interaction*. Addison-Wesley, 1987.

[213] A. Skowron and L. Polkowski. Synthesis of decision systems from data tables. In T.Y. Lin and N. Cercone, eds, *Rough Sets and Data Mining. Analysis of Imprecise Data*, pp. 259–299. Kluwer Academic Publishers, 1997.

[214] G. Snelting. Reengineering of configurations based on mathematical concept analysis. *ACM Transactions on Software Engineering and Methodology*, 5(2):146–189, 1996.

[215] G. Snelting and F. Tip. Reengineering class hierarchies using concept analysis. In *Proceedings of ACM SIGSOFT 6th International Symposium on Foundations of Software Engineering*, pp. 99–110, Lake Buena Vista, FL, USA, 1998.

[216] D. Soergel. Mathematical analysis of documentation systems. *Information Storage and Retrieval*, 3:129–173, 1967.

[217] J. Sowa. *Conceptual Structures: Information Processing in Mind and Machine*. Addison-Wesley, 1984.

[218] A. Spink and T. Saracevic. Interaction in information retrieval: selection and effectiveness of search terms. *Journal of the American Society for Information Science*, 48(8):741–761, 1997.

[219] A. Spoerri. InfoCrystal: Integrating exact and partial matching approaches through visualization. In *Proceedings of RIAO 94: Intelligent Multimedia Information Retrieval Systems and Management*, pp. 687–696, New York, USA, 1994.

[220] J. Stasko and E. Zhang. Focus+context display and navigation techniques for enhancing radial, space-filling hierarchy visualizations. In *Proceedings of IEEE Symposium on InfoVis 2000*, pp. 57–65, Salt Lake City, UT, USA, 2000.

[221] R. Stepp and R.S. Michalski. Conceptual clustering: inventing goal-oriented classifications of structured objects. In R.S. Michalski, J.G. Carbonell, and T.M. Mitchell, eds, *Machine Learning: An Artificial Intelligence Approach* (Vol. 2), pp. 471–498. Morgan Kaufmann, 1986.

[222] G. Stumme. Exploration tools in formal concept analysis. In E. Diday, Y. Lechevallier, and O. Opitz, eds, *Ordinal and Symbolic Data analysis.*

Proceedings of the International Conference on Ordinal and Symbolic Data Analysis (Osda 95), pp. 31–44. Springer, 1996.

[223] G. Stumme. Concept exploration–knowledge discovery in conceptual knowledge systems. PhD thesis, TU Darmstadt, Germany, 1997.

[224] G. Stumme. Local scaling in conceptual data systems. In *Proceedings of the 6th International Conference on Conceptual Structures*, pp. 308–320, Montpellier, France, 1998.

[225] G. Stumme. Conceptual knowledge discovery with frequent concept lattices. Technical Report FB4–Preprint No. 2043, TU Darmstadt, Germany, 1999.

[226] G. Stumme. Efficient data mining based on formal concept analysis. In *Proceedings of the 13th International Conference on Database and Expert Systems Applications (DEXA 2002)*, pp. 534–546, Aix-en-Provence, France, 2002.

[227] G. Stumme and A. Mädche. FCA-Merge: Bottom-up merging of ontologies. In *Proceedings of the 17th International Conference on Artificial Intelligence*, pp. 225–230, Seattle, WA, 2001.

[228] G. Stumme, R. Taouil, Y. Bastide, N. Pasquier, and L. Lakhal. Intelligent structuring and reducing of association rules with formal concept analysis. In F. Baader, G. Brewker, and T. Eiter, eds, *KI 2001: Advances in Artificial Intelligence*, pp. 335–350. Springer, 2001.

[229] G. Stumme, R. Taouil, Y. Bastide, N. Pasquier, and L. Lakhal. Computing iceberg concept lattices with Titanic. *Journal on Knowledge and Data Engineering*, 42(2):189–222, 2002.

[230] G. Stumme, R. Taouil, Y. Bastide, N. Pasquier, and R. Lakhal. Fast computation of concept lattices using data mining techniques. In *Proceedings of the 7th International Workshop on Knowledge Representation meets Databases*, pp. 129–139, Berlin, Germany, 2000.

[231] G. Stumme and R. Wille. A geometrical heuristic for drawing concept lattices. In R. Tamassia and I.G. Tollis, eds, *Graph Drawing*, pp. 452–459. Springer, Berlin, 1995.

[232] S. Sweeney, F. Crestani, and A. Tombros. Mobile delivery of news using hierarchical query-biased summaries. In *Proceedings of ACM SAC 2002, ACM Symposium on Applied Computing*, pp. 634–639, Madrid, Spain, 2002.

[233] L. Talavera and J. Bejar. Integrating declarative knowledge in hierarchical clustering tasks. In *Proceedings of the International Symposium on Intelligent Data Analysis*, pp. 211–222, Amsterdam, The Netherlands, 1999.

[234] R. Taouil and Y. Bastide. Computing proper implications. In *Proceedings of the ICCS-2001 International Workshop on Concept Lattice-Based Theory, Methods and Tools for Knowledge Discovery in Databases*, pp. 49–61, Palo Alto, CA, USA, 2001.

[235] L. Terveen and W. Hill. Finding and visualizing intersite clan graphs. In *Proceedings of ACM CHI'98: Human Factors in Computing Systems*, pp. 448–455, Los Angeles, 1998.

[236] K. Thompson, P. Langley, and W. Iba. Using background knowledge in concept formation. In *Proceedings of the 8th International Conference on Machine Learning*, pp. 554–558, Evanston, IL, USA, 1991.

[237] H. Toivonen, M. Klemettinen, P. Ronkainen, K. Haetoenen, and H. Mannila. Pruning and grouping of discovered association rules. In *Working Notes of the MLnet Workshop on Statistics, Machine Learning and Knowledge Discovery in Databases*, pp. 47–52, Heraklion, Crete, Greece, 1995.

[238] A. Tombros and M. Sanderson. Advantages of query biased summaries in information retrieval. In *Proceedings of the 21st Annual International ACM SIGIR Conference on Research and Development in Information Retrieval*, pp. 3–10, Melbourne, Australia, 1998.

[239] A. Tombros, R. Villa, and C. J. van Rijsbergen. The effectiveness of query-specific hierarchic clustering in information retrieval. *Information Processing & Management*, 38(4):559–582, 2002.

[240] J. Ullman. *Principles of Database and Knowledge-Base Systems, Vol. I.* Computer Science Press, 1988.

[241] P. Valtchev and R. Missaoui. Building concept (Galois) lattices from parts: generalizing the incremental methods. In *Proceedings of the 9th International Conference on Conceptual Structures*, pp. 290–303, Stanford, CA, USA, February 2001.

[242] P. Valtchev, R. Missaoui, R. Godin, and M. Meridji. Generating frequent itemsets incrementally: two novel approaches based on Galois lattice theory. *Journal of Experimental and Theoretical Artificial Intelligence*, 14(2–3):115–142, 2002.

[243] P. Valtchev, R. Missaoui, and P. Lebrun. A partition-based approach towards building Galois (concept) lattices. *Discrete Mathematics*, 256(3):801–829, 2002.

[244] F.J. van der Merwe and D.G. Kourie. Compressed pseudo-lattices. *Journal of Experimental and Theoretical Artificial Intelligence*, 14(2–3):229–254, 2002.

[245] C. J. van Rijsbergen. *Information Retrieval*. Butterworths, 1979.

[246] A. Veerasamy and R. Heikes. Effectiveness of a graphical display of retrieval results. In *Proceedings of the 20th Annual International ACM SIGIR Conference on Research and Development in Information Retrieval*, pp. 236–245, Philadelphia, USA, 1997.

[247] F. Vogt, C. Wachter, and R. Wille. Data analysis based on a conceptual file. In H.-H. Bock, W. Lenski, and P. Ihm, eds, *Classification, Data Analysis and Knowledge Organization*, pp. 131–140. Springer, 1991.

[248] F. Vogt and R. Wille. TOSCANA–A graphical tool for analyzing and exploring data. In R. Tamassia and I.G. Tollis, eds, *Graph Drawing'94*, pp. 226–233. Springer, 1995.

[249] E.M. Voorhees. Using Wordnet to disambiguate word senses for text retrieval. In *Proceedings of the 16th Annual International ACM SIGIR Conference on Research and Development in Information Retrieval*, pp. 171–180, Pittsburgh, PA, USA, 1993.

[250] K. Wagstaff, C. Cardie, S. Rogers, and S. Schroedl. Constrained k-means clustering with background knowledge. In *Proceedings of the 18th International Conference on Machine Learning*, pp. 577–584, Williamstown, MA, USA, 2001.

[251] K. Wang, Y. He, D. W. Cheung, and F.Y.L. Chin. Mining confident rules without support requirement. In *Proceedings of the 10th International ACM Conference on Information and Knowledge Management (CIKM 2001)*, pp. 89–96, Atlanta, GA, USA, 2001.

[252] G. Webb. OPUS: An efficient admissible algorithm for unordered search. *Journal of Artificial Intelligence Research*, 3:431–465, 1995.

[253] R. Wille. Restructuring lattice theory: an approach based on hierarchies of concepts. In I. Rival, ed., *Ordered Sets*, pp. 445–470. Reidel, 1982.

[254] R. Wille. Subdirect decomposition of concept lattices. *Algebra Universalis*, 17:275–287, 1983.

[255] R. Wille. Line diagrams of hierarchical concept systems. *International Classification*, 11(2):77–86, 1984.

[256] R. Wille. Complete tolerance relations of concept lattices. In G. Eigenthailer, H.K. Kaiser, W.B. Müller, and W. Nöbauer, eds, *Contributions to General Algebra*, Volume 3, pp. 397–415. Hölder-Pichler-Tempsky, Vienna, 1985.

[257] R. Wille. Tensorial decomposition of concept lattices. *Order*, 2:81–95, 1985.

[258] R. Wille. Subdirect product construction of concept lattices. *Discrete Mathematics*, 63:305–313, 1987.

[259] R. Wille. Knowledge acquisition by methods of formal concept analysis. In E. Diday, ed., *Data Analysis, Learning Symbolic and Numeric Knowledge*, pp. 365–380. Nova Science Publishers, 1989.

[260] R. Wille. Lattices in data analysis: how to draw them with a computer. In I. Rival, ed., *Algorithms and Order*, pp. 33–58. Kluwer, 1989.

[261] R. Wille. Concept lattices and conceptual knowledge systems. *Computers & Mathematics with Applications*, 23:493–515, 1992.

[262] R. Wille. The basic theorem of triadic concept analysis. *Order*, 12:149–158, 1995.

[263] R. Wille. Conceptual graphs and formal concept analysis. In *Proceedings of the 5th International Conference on Conceptual Structures*, pp. 290–303, Seattle, WA, 1997.

[264] R. Wille. Conceptual landscapes of knowledge: A pragmatic paradigm for knowledge processing. In H. Gaul and H. Locarek-Junge, eds, *Classification in the Information Age*, pp. 344–356. Springer, 1999.

[265] P. Willet. Recent trends in hierarchic document clustering: a critical review. *Information Processing & Management*, 24(5):577–597, 1988.

[266] G. Wills. NicheWorks–interactive visualization of very large graphs. *Journal of Computational and Graphical Statistics*, 8(2):190–212, 1999.

[267] I.H. Witten and E. Frank. *Data Mining: Practical Machine Learning Tools and Techniques with Java Implementations*. Morgan Kaufmann, 2000.

[268] S.K.M. Wong, W. Ziarko, V.V. Raghavan, and P.C.N. Wong. On modeling of information retrieval concepts in vector spaces. *ACM Transactions on Database Systems*, 12(2):299–321, 1987.

[269] Z. Xie, W. Hsu, Z. Liu, and M.L. Lee. Concept lattice based composite classifiers for high predictability. *Journal of Experimental and Theoretical Artificial Intelligence*, 14(2–3):143–156, 2002.

[270] J. Xu and W.B. Croft. Query expansion using local and global document analysis. In *Proceedings of the 19th Annual International ACM SIGIR Conference on Research and Development in Information Retrieval*, pp. 4–11, Zurich, Switzerland, 1996.

[271] J. Xu and W.B. Croft. Improving the effectiveness of information retrieval with local context analysis. *ACM Transactions on Information Systems*, 18(1):79–112, 2000.

[272] C. Yu, W. Meng, and S. Park. A framework for effective retrieval. *ACM Transactions on Database Systems*, 14(2):147–167, 1989.

[273] M. J. Zaki. Generating non-redundant association rules. In *Proceedings of the 6th ACM SIGKDD International Conference on Knowledge Discovery and Data Mining*, pp. 34–43, Boston, USA, 2000.

[274] M.J. Zaki and C.-J. Hsiao. An efficient algorithm for closed association rule mining. Technical Report 99-10, Computer Science Dept., Rensselaer Polytechnic Institute, Troy, NY, USA, 1999.

[275] M.J. Zaki and C.-J. Hsiao. CHARM: An efficient algorithm for closed itemset mining. In *Proceedings of the 2nd SIAM International Conference on Data Mining*, Arlington, VA, USA, 2002.

[276] M.J. Zaki and M. Ogihara. Theoretical foundations of association rules. In *Proceedings of the 3rd SIGMOD'98 Workshop on Research Issues in Data Mining and Knowledge Discovery*, pp. 71–78, Seattle, WA, USA, 1998.

[277] O. Zamir and O. Etzioni. Grouper: A dynamic clustering interface to web search results. *WWW8/Computer Networks*, 31(11–16):1361–1374, 1999.

[278] C. Zhai and J. Lafferty. A study of smoothing methods for language models applied to ad hoc information retrieval. In *Proceedings of the 24th Annual International ACM SIGIR Conference on Research and Development in Information Retrieval*, pp. 334–342, New Orleans, LA, USA, 2001.

[279] W. Ziarko. A method for computing all maximally general rules in attribute-value systems. *Computational Intelligence*, 12(2):223–234, 1996.

Index

Concept Data Analysis: Theory and Applications. Claudio Carpineto and Giovanni Romano

ISBN: 0-470-85055-8